SpringerBriefs in Mathematics

Series editors

Nicola Bellomo
Michele Benzi
Palle Jorgensen
Tatsien Li
Roderick Melnik
Otmar Scherzer
Benjamin Steinberg
Lothar Reichel
Yuri Tschinkel
George Yin
Ping Zhang

SpringerBriefs in Mathematics showcases expositions in all areas of mathematics and applied mathematics. Manuscripts presenting new results or a single new result in a classical field, new field, or an emerging topic, applications, or bridges between new results and already published works, are encouraged. The series is intended for mathematicians and applied mathematicians.

More information about this series at http://www.springer.com/series/10030

Jianhai Bao · George Yin
Chenggui Yuan

Asymptotic Analysis for Functional Stochastic Differential Equations

Springer

Jianhai Bao
Department of Mathematics
Central South University
Changsha, Hunan
China

Chenggui Yuan
Department Mathematics
Swansea University
Swansea
UK

George Yin
Department of Mathematics
Wayne State University
Detroit, MI
USA

ISSN 2191-8198 ISSN 2191-8201 (electronic)
SpringerBriefs in Mathematics
ISBN 978-3-319-46978-2 ISBN 978-3-319-46979-9 (eBook)
DOI 10.1007/978-3-319-46979-9

Library of Congress Control Number: 2016953213

Mathematics Subject Classification (2010): 60H10, 60H15, 60J25, 60H30, 60J25, 60F10, 39B82

Printed on acid-free paper

This Springer imprint is published by Springer Nature
The registered company is Springer International Publishing AG
The registered company address is: Gewerbestrasse 11, 6330 Cham, Switzerland

To my father Chengzhu Bao and my mother Junhua Han

Jianhai Bao

To my wife Meimei Zhang

George Yin

To my wife Wangdong Yang

Chenggui Yuan

Preface and Introduction

This work focuses on dynamical systems that involve delays and random disturbances. The motivation of our study stems from a wide variety of systems in real life in which random noise has to be taken into consideration and the effect of delays cannot be avoided or ignored. In fact, there are numerous sources of delays that we encounter in every-day life. For example, everyone has the experience of delays in highway traffic, air traffic, and package delivery etc. Accompanying the rapid progress in technology, digital computers, internet, and various hand-held "smart" devices, delays become more and more prevalent in physical, cyber-physical, and biological systems. In the new era, delays are ubiquitous in wired or wireless communications, queueing, and networked control problems.

In this work, we concentrate on such systems that are described by functional stochastic differential equations. A functional stochastic differential equation is one whose state space depends on the past. Accompanying the extensive literature on functional differential equations for deterministic systems (see, e.g., the monograph [62]), the study of such stochastic systems was initiated in 1964 by Itô and Nisio [69]. For regularity and stochastic stability of solution processes, and Markov trajectories plus the infinitesimal generator for segment processes, we refer to the monographs [96] and [106], respectively. Because their importance, significant efforts have also been devoted to the study of delay differential systems and more generally functional differential systems, which are exemplified by chemical processes, biological systems, and communication systems. It is observed that delays may have detrimental impact on stability and performance of dynamical systems. Their presence adds substantial difficulties in analysis of long-term behavior of systems and stability. Facing the challenges, we need to have a thorough understanding on systems with delays and treat such systems with great care.

Motivational Examples

There are many examples and numerous applications involving functional differential equations with noise. Among the early works on controlled stochastic differential delay equations is [85], whereas early treatment of stability of stochastic delay equations can be found in [82]. In recent years, emerging applications in financial engineering have drawn much attention; see [2, 4, 7, 75] and references therein. Although Black-Scholes formula is one of the most important results in finance, the feasibility of the model has been questioned because of the assumption of constant appreciation and volatility rates. To treat more realistic situation, effort has also been directed to the development of dynamical models that take into consideration the influence of past history.

This brief is devoted to the study of large time behavior of stochastic functional differential equations. As a motivational example, we consider the following controlled functional stochastic differential equations in an infinite horizon, which is a modification of that considered in [84, Chapter 5]. Suppose that $X(t)$ is a diffusion process living in $\mathbb{R}^n$ and that X_t is the associated memory segment process $X_t = \{X(t+\theta) : -\tau \le \theta \le 0\}$ with $\tau > 0$. Let U be the set of controls, and $b(\cdot)$ and $\sigma(\cdot)$ be appropriate functions (more precise notation will be given in later chapters), and $W(t)$ be a standard Brownian motion. Consider the following functional stochastic differential equation

$$\mathrm{d}X(t) = b(X_t, u(X_t))\mathrm{d}t + \sigma(X_t)\mathrm{d}W(t).$$

We wish to minimize a long run average cost functional given by

$$\lim_{T\to\infty} \frac{1}{T}\mathbb{E}^u\left[\int_0^T k(X_t, u(X_t))\, dt\right].$$

The idea of the ergodic cost problem is to replace the instantaneous measure by that of the ergodic measure so that the expectation becomes an average with respect to the ergodic measure leading to a substantial reduction of complexity. Such study is of great utility in many applications. However, to be able to carry out the desired task, one needs to make sure that under suitable conditions, there is an ergodic measure for such systems. This is one of the main objectives of the current brief.

As a second example, consider an adaptive control problem of controlled diffusion with past dependence treated in [46]. For a fixed $\tau > 0$, let $\mathscr{C} = C([-\tau, 0]; \mathbb{R}^n)$ be the space of continuous functions. Consider a controlled process defined by

$$\begin{cases} \mathrm{d}X(t) = \{f_1(X(t)) + f_2(X_t, u(X_t), \alpha^*)\}\mathrm{d}t + g(X(t))\mathrm{d}W(t) \\ X(0) = \xi \in \mathscr{C}, \end{cases}$$

where $X(t) \in \mathbb{R}^n$, α^* is an unknown parameter living in $\mathbb{R}^d$, and $W(\cdot)$ is a standard $\mathbb{R}^n$-valued Brownian motion, $u(\cdot) \in U \subset \mathbb{R}^m$ with U being a compact set and $X_t = \{X(t+\theta) : -\tau \leq \theta \leq 0\}$. The functions $f_1(\cdot)$ and $g(\cdot)$ satisfy global Lipschitz condition. The diffusion is assumed to be non-degenerate in that $g(x)g^*(x) \geq cI_{n\times n}$ for some $c > 0$ and all $x \in \mathbb{R}^n$ with g^* denoting the transpose of g. The function $f_2(\cdot)$ is assumed to be bounded and Borel measurable on $\mathscr{C}$. The objective is to minimize the long-run average cost

$$J(u,\xi,\alpha^*) = \limsup_{T\to\infty} \frac{1}{T} E_{\xi,u,\alpha^*} \int_0^T k(X_s, u(X_s))ds,$$

where $k(\cdot) : \mathscr{C} \times U \mapsto \mathbb{R}$ is a bounded and continuous function. In [46], it was shown that the unknown parameter α^* can be estimated by a biased maximum likelihood estimator and a certainty equivalent control can be obtained. The control so designed yields an almost self-optimizing adaptive control. Again, in this process, the ergodicity plays an essential role.

Recently, there are much interests in treating the so-called consensus formation for networked systems and multi-agent systems. Related published works in physics, computer science, control engineering, ecology, biology, social sciences, as well as in the journal Nature among others, show the growing interests from a wide range of communities. The goal is to achieve a common objective such as position, speed, load distribution, etc. for a group of members often referred to as mobile agents. Such models have been successfully used since late 1980s in computer graphics, physics, control engineering, and social network modeling, to describe collective behavior such as flocking, schooling, autonomous vehicles, and other group behavior among others. A discrete-time model of autonomous agents, which can be viewed as points or particles, all moving in the same plane with the same speed but with different headings was proposed in [132], in which each agent updates its heading using a local rule based on the average headings of its own and its neighbors. This model turns out to be related to a formulation introduced earlier for simulating animation of flocking and schooling behaviors in [120]. A version of the scheme taking into consideration of time delay with communication latency and possibly random-switching topology may be written as a stochastic approximation type algorithm of the following form

$$x_{n+1} = x_n + \mu M(\alpha_n) x_{n-\lfloor d/\mu \rfloor} + \mu \hat{G}(x_{n-\lfloor d/\mu \rfloor}, \alpha_n, \zeta_{n-\lfloor d/\mu \rfloor}, \tilde{\zeta}_n),$$

with the initial segment x_k for $k = -\lfloor d/\mu \rfloor, \ldots, 0$ being arbitrary. In the above, x_n is in a multi-dimensional Euclidean space, $\mu > 0$ is the stepsize of consensus control algorithm, $\hat{G}$ is an appropriate function, $d > 0$ is a constant and $\lfloor d/\mu \rfloor$ denotes the integer part of d/μ representing the time delays, α_n is a discrete-time Markov chain with state space $\mathscr{M} = \{1, \ldots, m_0\}$ and transition matrix $P^\varepsilon = I + \varepsilon Q$ ($\varepsilon > 0$ is a small parameter, I is the identity matrix, and Q is a generator of a continuous-time Markov chain), $M(i)$ is the generator of a continuous-time Markov chain for each $i \in \mathcal{M}$, $d > 0$ is a constant with $\lfloor d/\mu \rfloor$ representing the delay, and $\{\zeta_n\}$ and $\{\tilde{\zeta}_n\}$ are sequences representing measurement or observation noise. Assuming that $\mu = O(\varepsilon)$ and taking a continuous-time interpolation $x^\varepsilon(t) = x_n$ and $\alpha^\varepsilon = \alpha_n$ for $t \in [n\varepsilon, (n+1)\varepsilon)$. Then $x^\varepsilon(\cdot)$ belongs to a function space that consists of functions that are right continuous with left limits endowed with the Skorohod topology. Under appropriate conditions, it can be shown that the noises are averaged out, $x^\varepsilon(\cdot)$ has a weak limit characterized by the switched pure delay equation $\dot{x}(t) = M(\alpha(t))x(t-d)$, where $\alpha(\cdot)$ is a continuous-time Markov chain [the weak limit of $\alpha^\varepsilon(\cdot)$]. Furthermore, it can be demonstrated that $x^\varepsilon(\cdot + t_\varepsilon)$ (with $t_\varepsilon \to \infty$ as $\varepsilon \to 0$) converges weakly to θ, a vector with all components being the same (i.e., consensus is reached). With more effort, one can analyze the corresponding estimation error sequence $\{x_n - \theta\}$ and show that a suitable scaling leads to a stochastic differential delay equation with Markov switching of the form

$$\mathrm{d}z(t) = M(\alpha(t))z(t-d)\mathrm{d}t + \overline{G}(\alpha(t))\mathrm{d}W(t),$$

for an appropriate function $\overline{G}(\cdot)$ and a standard multidimensional Brownian motion $W(\cdot)$; see [148] for more details together with many references therein.

Purposes of This Brief

This brief is mainly for research purposes. It is written for probabilists, applied mathematicians, engineers, and scientists who need to use delay systems in their work. Selected topics from the brief can also be used in a graduate level topics course in probability and stochastic processes.

The format of the SpringerBriefs in Mathematics gives us an excellent opportunity to focus on the long-term behavior of the functional stochastic differential equations. This very focused approach enables us to emphasize the central theme; our study encompasses ergodicity. Although there are many excellent treatises on stochastic differential delay equations and functional stochastic differential equations, short monographs devoted to ergodicity of functional stochastic differential equations seem to be scarce to date. It would certainly be welcomed to have a work collect a number of long-run properties about functional stochastic systems. Nevertheless, because of the format of the Briefs, we are not able to make this book a comprehensive treatment of functional stochastic differential equations. In fact, we are not even able to include the vast literature in a short book like this.

Outline of the Book

This book is organized as follows. Chapter 1 is devoted to ergodicity of functional stochastic differential equations and Chapter 2 focuses on ergodicity without dissipative conditions. Chapter 3 ascertains rates of convergence of Euler–Maruyama procedures. Chapter 4 obtains large deviations estimates for neutral functional stochastic differential equations with jumps. Chapter 5 gives an application to an interest model in an infinite horizon. Finally, to make the brief relatively self-contained, two appendices are given at the end of the book to collect a number of results on existence and uniqueness of solutions of functional stochastic differential equations, Markov properties, as well as certain technical results such as variation of constants formulas.

Acknowledgements

Without the help and encouragement of many people, this book project would not have been completed. We would like to use this opportunity to thank many colleagues and friends helping us in bringing this brief into being. In particular, Mu-Fa Chen, Xianping Guo, Zhenting Hou, Junhao Hu, Niels Jacob, Junping Li, Zaiming Liu, Xuerong Mao, Jinghai Shao, Aubrey Truman, Feng-Yu Wang, Le Yi Wang, Fuke Wu, Jiang-Lun Wu, and Qing Zhang worked with us on research problems for systems with delays, functional stochastic differential equations, and related issues. We thank them for their help, encouragement, and inspiration.

We thank the reviewers for the constructive comments and suggestions. Our thanks also go to Donna Chernyk and Springer professionals for their help and assistance in finalizing the book. The research of J. Bao was supported in part by the National Natural Science Foundation of China under grant No. 11401592, and the research of C. Yuan was supported in part by the EPSRC and NERC. During the past years, the research of G. Yin was supported in part by the National Science Foundation, the Air Force Office of Scientific Research, and the Army Research Office, under different research projects for different research topics during different times. The supports of the funding agencies are greatly appreciated.

Changsha, China
Detroit, USA
Swansea, UK
August 2016

Jianhai Bao
George Yin
Chenggui Yuan

Contents

Notations and Abbreviations

$\|A\|_{HS}$	Hilbert–Schmidt norm for an operator A
$\mathscr{B}(\mathscr{C})$	σ-algebra on $\mathscr{C}$
$\mathscr{C}$	$= C([-\tau,0];\mathbb{R}^n)$: collection of continuous functions
$C(I;\mathbb{R}^n)$	set of all continuous functions from I to $\mathbb{R}^n$
$\mathscr{D}$	$= D([-\tau,0];\mathbb{R}^n)$: collection of right continuous functions with left limits endowed with the Skorohod topology
$\mathbb{E}$	expectation
EM method	Euler–Maruyama method
FSDE	functional stochastic differential equation
FSPDE	functional stochastic partial differential equation
$(H, \langle\cdot,\cdot\rangle_H, \|\cdot\|_H)$	real separable Hilbert space
LDP	large deviations principle
$L(\mathscr{C})$	space of Lipschitz continuous functions
$\mathrm{Re}(z)$	real part of z
$\mathbb{R}^n$	n-dimensional Euclidean space
$\mathbb{R}^n \otimes \mathbb{R}^m$	collection of all $n \times m$ matrices
SDDE	stochastic differential delay equation
SPDE	stochastic partial differential equation
$W(t)$	m-dimensional standard Brownian motion
X_t	$= \{X(t+\theta) : -\tau \leq \theta \leq 0\}$
a^+	$= \max\{a,0\}$ for a real number a
a^-	$= \max\{-a,0\}$ for a real number a
$a \wedge b$	$= \min\{a,b\}$
$a \vee b$	$= \max\{a,b\}$
a.s.	almost surely
$a \lesssim b$	$a \leq cb$ for some $c > 0$
$a \lesssim_T b$	$a \leq cb$ with c depending on T
v^*	transpose of $v \in \mathbb{R}^{l_1 \times l_2}$ with $l_1, l_2 \geq 1$
$\|x\|$	Euclidean norm for $x \in \mathbb{R}^n$
$(\Omega, \mathscr{F}, (\mathscr{F}_t)_{t\geq 0}, \mathbb{P})$	filtered probability space

τ	$\in (0,)$: delay or time lag
$\|\zeta\|_\infty$	$:= \sup_{-\tau \le \theta \le 0} \|\zeta(\theta)\|$ for $\zeta \in \mathscr{C}$
∇	gradient
∇^2	Hessian
$\Rightarrow$	convergence in distribution

Chapter 1
Ergodicity for Functional Stochastic Equations Under Dissipativity

In this chapter, we investigate ergodicity for certain classes of functional stochastic equations including functional stochastic differential equations (FSDEs for short) driven by Brownian motions, FSDEs of neutral type, FSDEs driven by jump processes, functional stochastic partial differential equations (FSPDEs for abbreviation) driven by cylindrical Wiener processes, and FSPDEs driven by cylindrical α-stable processes. Sections 1.1–1.4 and 1.6 and 1.7 of this chapter are mainly based on the results from [12].

1.1 An Overview of FSDEs

More often than not, delays are unavoidable in a wide range of applications. In response to the great needs, there is an extensive literature on functional differential equations (see, e.g., the monographs [46, 65, 145]). An FSDE is an SDE whose state space depends on the past. Such SDE was initiated in 1964 by Itô and Nisio [72]. Some early studies on stability of stochastic delay equations can be found in [85]; recent consideration of randomly varying delay systems [144] generalizes the original work [153] on pure delay equations for deterministic systems. For regularity and stochastic stability of solution processes, and Markov trajectories plus the infinitesimal generator for segment processes, we refer to the monographs [99] and [109], respectively. In addition, in place of the Brownian motion, we may consider the associated wideband noise perturbations; see [150] and references therein.

Fix $\tau > 0$ and let $\mathscr{C} = C([-\tau, 0]; \mathbb{R}^n)$ be the collection of all continuous functions $f : \mathscr{C} \mapsto \mathbb{R}^n$ endowed with the uniform norm $\|f\|_\infty := \sup_{-\tau\leq\theta\leq0} |f(\theta)|$. In contrast to SDEs without memory, FSDEs enjoy numerous distinguished characteristics. To demonstrate those features, we consider an FSDE on $(\mathbb{R}^n, \langle\cdot,\cdot\rangle, |\cdot|)$ in the following form

$$\mathrm{d}X(t) = b(X_t)\mathrm{d}t + \sigma(X_t)\mathrm{d}W(t), \quad t > 0, \quad X_0 = \xi \in \mathscr{C}, \tag{1.1}$$

J. Bao et al., *Asymptotic Analysis for Functional Stochastic Differential Equations*, SpringerBriefs in Mathematics, DOI 10.1007/978-3-319-46979-9_1

where $X_t := \{X(t+\theta); \theta \in [-\tau, 0]\}$ is the segment (or the functional solution), $b : \mathscr{C} \mapsto \mathbb{R}^n$, $\sigma : \mathscr{C} \mapsto \mathbb{R}^n \otimes \mathbb{R}^m$ are Lipschitz continuous on each bounded set, and $(W(t))_{t\geq 0}$ is an m-dimensional Brownian motion defined on $(\Omega, \mathscr{F}, (\mathscr{F}_t)_{t\geq 0}, \mathbb{P})$, a filtered probability space. Let $\mathscr{B}_b(\mathscr{C})$ stand for the family of all bounded measurable functions $f : \mathscr{C} \mapsto \mathbb{R}$. Under appropriate conditions, (1.1) possesses the peculiarities below:

(i) $(X(t))_{t\geq -\tau}$ is not Markovian, whereas $(X_t)_{t\geq 0}$ is (strong) Markovian;
(ii) For each fixed $t \geq 0$, $X(t) \in \mathbb{R}^n$, while $X_t \in \mathscr{C}$;
(iii) Kolmogorov's backward equation is, in general, unavailable for the Markov transition semigroup generated by the segment process $(X_t)_{t\geq 0}$, i.e., $P_t f(\xi) := \mathbb{E} f(X_t(\xi))$, $f \in \mathscr{B}_b(\mathscr{C})$;
(iv) P_t is generally not strong Feller.

According to (i), we define the Markov transition semigroup P_t generated by X_t, not $X(t)$, i.e., $P_t f(\xi) = \mathbb{E} f(X_t(\xi))$, $f \in \mathscr{B}_b(\mathscr{C})$. From (ii), we know that the solution process $(X(t))_{t\geq -\tau}$ is finite-dimensional, nevertheless the corresponding segment process $(X_t)_{t\geq 0}$ is infinite-dimensional. As we know, for finite-dimensional diffusion processes, boundedness in probability implies existence of an invariant measure. However, for the infinite-dimensional cases, boundedness in probability no longer guarantees the implication [37, Chapter 6]. For diffusion processes without memory, Kolmogorov's backward equation is one of the effective tools to establish functional inequalities (see, e.g., the monograph [137]), whereas as far as FSDEs are concerned, in accordance with (iii) functional inequalities in general cannot be established by using Kolmogorov's backward equation, since, for a rather long period of time, there was no appropriate operator available for systems with delay and no Itô's formula was available. The Itô formula has only been obtained very recently by Dupire [50]. Subsequent work pointed out a viable alternative for future study; see [32, 52], and references therein.

By Doob's theorem [37, Theorem 4.2.1, p. 43], strong Feller and irreducibility imply that the associated Markov transition semigroup admits at most one invariant measure. While, for FSDEs, Doob's theorem does not work well to investigate uniqueness of invariant measure since the semigroup is generally not strong Feller; see, for instance, $\mathrm{d}X(t) = X(t-1)\mathrm{d}W(t)$ constructed in [128]. Moreover, the classical Itô formula cannot be applied to the functional of the segment process $(X_t)_{t\geq 0}$ due to the fact that $(X_t)_{t\geq 0}$ is not a semi-martingale. As infinite-dimensional Markov processes with finite-dimensional noises, the functional solutions are highly degenerated. From the analysis above, we find that a number of existing tools are no longer applicable to FSDEs, in particular, for the corresponding segment processes. To an extent, behavior of FSDEs may be markedly different from diffusion processes without memory.

1.2 Introduction

The ergodicity of SDEs and SPDEs, in which the state spaces are independent of the past, has been studied extensively. So far, there are several approaches to investigate ergodicity for finite or infinite-dimensional stochastic dynamical systems; see, for instance, [105, 122] using the Lyapunov function argument (see, e.g., Meyn-Tweedie [107]), [47, 114] using Harris' theorem (see, e.g., [105, Theorem 1.5]), and [139, 157] using the coupling method. Further references on ergodicity of infinite-dimensional systems can be found in the monograph [37] and the lecture notes [63].

For FSDEs, one of the classical methods for showing existence of an invariant measure is to exhibit an accumulation point of a sequence of Krylov–Boguliubov measures (see, e.g., [37, Theorem 3.1.1, p. 21]) by using the tightness criterion of probability measures on the continuous function space (see, e.g., [20, Theorem 8.5, p. 55]). For existence of invariant measures, Es-Sarhir et al. [54] and Kinnally-Williams [79] considered FSDEs with superlinear drift term and positivity constraints, respectively. Recently, by an asymptotic coupling approach, Hairer et al. [64] addressed the open problem on uniqueness of invariant measures for a class of nondegenerate FSDEs. However, conditions imposed therein may not ensure existence of an invariant measure; see [64, Remark 3.2, p. 237]. Butkovsky [27] studied the rate of convergence to an invariant measure for Markov processes using the Wasserstein metric, and applied the result to SDDEs to obtain subgeometric ergodicity of the strong solutions. For the ergodicity of stochastic evolution equations (SEEs), Komorowski et al. [83] introduced the theories of weak-* ergodicity and e-property, these theories are then applied to study the invariant measure of SEEs.

Although the approach adopted in [54, 79] proved to be successful on investigating existence of invariant measures for a wide range of FSDEs, such a method does not work well for FSDEs of neutral type, FSDEs driven by jump processes and FSPDEs. In this chapter, under certain dissipative conditions, we adopt the remote start method, which is also called the dissipativity method and will be expounded in the proof of Theorem 1.5 below, to establish existence and uniqueness of invariant measures for several kinds of functional stochastic equations, which cover FSDEs driven by Brownian motions, FSDEs of neutral type, FSDEs driven by jump processes, FSPDEs driven by cylindrical Wiener processes, and FSPDEs driven by cylindrical α-stable processes. The key idea here is that in lieu of with the current time, we start the procedure from remote past.

1.3 Ergodicity for FSDEs

Throughout this section, we assume that $b(\cdot)$ and $\sigma(\cdot)$ are locally Lipschitz, and the initial value $X_0 = \xi \in \mathscr{C}$ is independent of $(W(t))_{t\geq 0}$. Let $\rho(\cdot)$ be a probability measure on $[-\tau, 0]$. For any $\phi, \psi \in \mathscr{C}$, we further assume the following conditions.

(H1) There exist constants $\lambda_1 > \lambda_2 > 0$ such that

$$2\langle\phi(0)-\psi(0), b(\phi)-b(\psi)\rangle + \|\sigma(\phi)-\sigma(\psi)\|_{HS}^2 \\ \leq -\lambda_1|\phi(0)-\psi(0)|^2 + \lambda_2\int_{[-\tau,0]}|\phi(\theta)-\psi(\theta)|^2\rho(\mathrm{d}\theta),$$

where $\|\cdot\|_{HS}$ denotes the Hilbert–Schmidt norm.

(H2) There exists a constant $\lambda_3 > 0$ such that

$$\|\sigma(\phi)-\sigma(\psi)\|_{HS}^2 \leq \lambda_3\Big(|\phi(0)-\psi(0)|^2 + \int_{[-\tau,0]}|\phi(\theta)-\psi(\theta)|^2\rho(\mathrm{d}\theta)\Big).$$

Remark 1.1 Although we do not need the full capacity of assumption (H1) for the well-posedness of (1.1), we use (H1) to explore the existence and uniqueness of invariant measure for $(X_t(\xi))_{t\geq 0}$. The well-posedness of (1.1) can indeed be guaranteed by the following coercivity condition

$$2\langle\phi(0), b(\phi)\rangle + \|\sigma(\phi)\|_{HS}^2 \lesssim 1 + |\phi(0)|^2 + \int_{[-\tau,0]}|\phi(\theta)|^2\rho(\mathrm{d}\theta), \quad \phi\in\mathscr{C}. \quad (1.2)$$

For instance, $b(\phi) = -\phi(0)^3 - \phi(0) + \int_{-1}^0\phi(\theta)\mathrm{d}\theta$, and $\sigma(\phi) = \phi(0)^2$ for $\phi \in \mathscr{C} := C([-1,0];\mathbb{R})$ such that (1.2) holds. In fact, since b and σ are locally Lipschitz continuous, by the standard truncation procedure (see, e.g., [99, p. 57]), (1.1) admits a unique strong solution $(X(t;\xi))_{t\geq-\tau}$ with the initial datum $X_0 = \xi \in \mathscr{C}$ up to the life time $\tau_\infty := \lim_{n\to\infty}\tau_n$ (see, e.g., [99, Theorem 2.8, p. 154]), where $\tau_n := \inf\{t > 0 : |X(t)| \geq n\}$ for any integer $n > \|\xi\|_\infty$. By Hölder's inequality and Itô's isometry, we deduce from (1.2) that, for any $T > 0$, there exists $C_T > 0$ independent of n such that

$$C_T \geq \mathbb{E}|X(T\wedge\tau_n;\xi)|^2 \geq \mathbb{E}|X(T\wedge\tau_n;\xi)|^2 I_{\{\tau_n\leq T\}} \geq n^2\mathbb{P}(\tau_n\leq T).$$

This further gives that

$$\mathbb{P}(\tau_n \leq T) \leq C_T/n^2.$$

So, one has $\sum_{n=1}^\infty \mathbb{P}(\tau_n\leq T) < \infty$. Then, by the Borel–Cantelli lemma (see, e.g., [99, Lemma 2.4, p. 7]), we conclude that $\mathbb{P}(\tau_\infty \leq T) = 0$. Due to the arbitrariness of T, we arrive at $\mathbb{P}(\tau_\infty = \infty) = 1$. Therefore, (1.1) is well-posed. Therefore, under (H1), (1.1) admits a unique strong solution $(X(t;\xi))_{t\geq-\tau}$; see Appendix A for more details on the existence and uniqueness.

By virtue of Theorem B.1 (see also [109, Theorem 1.1, p. 51]), we can show the segment process $(X_t(\xi))_{t\geq 0}$ is a Markov process. Define the Markovian transition semigroup $(P_t)_{t\geq 0}$ associated with the segment process $(X_t(\xi))_{t\geq 0}$ by

$$P_t f(\xi) = \mathbb{E} f(X_t(\xi)), \quad t \ge 0, \ \xi \in \mathscr{C}, \ f \in \mathscr{B}_b(\mathscr{C}).$$

By the Markov property, we have $P_t \circ P_s = P_{t+s}$, $t, s \ge 0$, where $P_t \circ P_s$ means the composition of the operators P_t and P_s.

Definition 1.2 A probability measure $\mu \in \mathscr{P}(\mathscr{C})$, the collection of all probability measures on $\mathscr{C}$, is invariant w.r.t. $(P_t)_{t\ge 0}$ if

$$\mu(P_t f) = \mu(f) := \int_{\mathscr{C}} f(\xi)\mu(\mathrm{d}\xi), \quad t \ge 0, \ f \in \mathscr{B}_b(\mathscr{C}).$$

In this case, (1.1) is said to admit an invariant measure.

Remark 1.3 If $\mu \in \mathscr{P}(\mathscr{C})$ is an invariant measure of (1.1) and $\mathscr{L}(\xi) = \mu$, where $\mathscr{L}(\xi)$ means the law of ξ, by [3, Lemma 1.1.9, p. 14], the independence of $\xi \in \mathscr{C}$ and $(W(t))_{t\ge 0}$, and the double law property of conditional expectation, one has

$$\mu(f) = \int_{\mathscr{C}} \mathbb{E} f(X_t(\eta))\mu(\mathrm{d}\eta) = \mathbb{E}(\mathbb{E}(f(X_t(\xi)))|\mathscr{F}_0) = \mathbb{E}(f(X_t(\xi))).$$

Thus, we conclude $\mathscr{L}(X_t(\xi)) = \mu$, i.e., the law of $X_t(\xi)$ is invariant under time translation.

Definition 1.4 An invariant measure μ for $(P_t)_{t\ge 0}$ is said to be exponentially mixing with exponent $\lambda > 0$ and function $c : \mathscr{C} \mapsto (0, \infty)$ if

$$|P_t f(\xi) - \mu(f)| \le c(\xi) e^{-\lambda t} \|f\|_{\mathrm{Lip}}, \quad t \ge 0, \ \xi \in \mathscr{C}, \ f \in L(\mathscr{C}),$$

in which $L(\mathscr{C})$ denotes the set of all Lipschitz functions $f : \mathscr{C} \mapsto \mathbb{R}$, and $\|f\|_{\mathrm{Lip}} := \sup_{\xi\ne\eta} \frac{|f(\xi)-f(\eta)|}{\|\xi-\eta\|_\infty}$, i.e., the Lipschitz constant of f. That is, the Markovian transition semigroup $(P_t)_{t\ge 0}$ converges exponentially fast to the equilibrium.

The following assertion shows that (1.1) admits a unique invariant measure.

Theorem 1.5 *Assume that* (H1) *and* (H2) *hold. Then,* (1.1) *has a unique invariant measure* $\mu \in \mathscr{P}(\mathscr{C})$*, which is exponentially mixing.*

Proof The entire proof of this theorem is divided into the following three steps.
Step 1: Existence of an invariant measure. We adopt a remote start method to show existence of an invariant measure of (1.1). Such an approach has been applied successfully in, e.g., Da Prato-Zabczyk [37, p. 105–108] and Prévôt-Röckner [113, p. 98–110], for obtaining existence of invariant measures for SPDEs under dissipative conditions.

Let $(\widetilde{W}(t))_{t\ge 0}$ be an independent copy of $(W(t))_{t\ge 0}$ defined on the probability space $(\Omega, \mathscr{F}, \mathbb{P})$. Define a double-sided Wiener process $(\overline{W}(t))_{t\in\mathbb{R}}$ by

$$\overline{W}(t) = \begin{cases} W(t), & t \geq 0, \\ \widetilde{W}(-t), & t < 0 \end{cases} \tag{1.3}$$

with the filtration

$$\overline{\mathscr{F}}_t := \bigcap_{s>t} \bar{\mathscr{F}}_s^0,$$

where $\overline{\mathscr{F}}_s^0 := \sigma(\{\overline{W}(r_2) - \overline{W}(r_1) : -\infty < r_1 \leq r_2 \leq s\}, \mathscr{N})$ with $\mathscr{N} := \{A \in \mathscr{F} | \mathbb{P}(A) = 0\}$. Following the argument of [113, Proposition 2.1.13, p. 16], we deduce that $\overline{W}(t) - \overline{W}(s)$ is independent of $\overline{\mathscr{F}}_s$ for any $-\infty < s < t < \infty$. Fix $s \in \mathbb{R}$ and consider the following FSDE

$$\mathrm{d}X(t) = b(X_t)\mathrm{d}t + \sigma(X_t)\mathrm{d}\overline{W}(t), \quad t \geq s, \quad X_s = \xi \in \mathscr{C}. \tag{1.4}$$

Under (H1), (1.4) admits a unique strong solution $X(t; s, \xi)$ with the initial data $X_s = \xi$ (see Remark 1.1). In what follows, let $-\infty < s_1 \leq s_2 \leq t < \infty$. It is easy to see that

$$\begin{aligned} X(t; s_1, \xi) - X(t; s_2, \xi) = X(s_2; s_1, \xi) - \xi(0) + \int_{s_2}^{t} \{b(X_u(s_1, \xi)) - b(X_u(s_2, \xi))\}\mathrm{d}u \\ + \int_{s_2}^{t} \{\sigma(X_u(s_1, \xi)) - \sigma(X_u(s_2, \xi))\}\mathrm{d}\overline{W}(u). \end{aligned}$$

For notational simplicity, set

$$\Gamma(t) := X(t; s_1, \xi) - X(t; s_2, \xi). \tag{1.5}$$

For any $\lambda > 0$, by the Itô formula and (H1), it follows that

$$\begin{aligned} e^{\lambda t}\mathbb{E}|\Gamma(t)|^2 \leq e^{\lambda s_2}\mathbb{E}|\Gamma(s_2)|^2 + \lambda_2 e^{\lambda\tau} \int_{s_2-\tau}^{s_2} e^{\lambda u}\mathbb{E}|X(u; s_1, \xi) - \xi(u - s_2)|^2\mathrm{d}u \\ - (\lambda_1 - \lambda - \lambda_2 e^{\lambda\tau}) \int_{s_2}^{t} e^{\lambda u}\mathbb{E}|\Gamma(u)|^2\mathrm{d}u. \end{aligned}$$

A simple consequence of $\lambda_1 > \lambda_2 > 0$ is that there exists a unique $\lambda > 0$ such that $\lambda_1 - \lambda - \lambda_2 e^{\lambda\tau} = 0$. Hence, we have

$$\begin{aligned} \mathbb{E}|\Gamma(t)|^2 \leq e^{-\lambda t}\Big\{e^{\lambda s_2}\mathbb{E}|\Gamma(s_2)|^2 \\ + \lambda_2 e^{\lambda\tau} \int_{s_2-\tau}^{s_2} e^{\lambda u}\mathbb{E}|X(u; s_1, \xi) - \xi(u - s_2)|^2\mathrm{d}u\Big\}. \end{aligned} \tag{1.6}$$

Note from (H1) and (H2) that there exist constants $\nu_1 > \nu_2 > 0$ such that

$$2\langle\psi(0), b(\psi)\rangle + \|\sigma(\psi)\|_{HS}^2 \leq c - \nu_1|\psi(0)|^2 + \nu_2 \int_{[-\tau,0]} |\psi(\theta)|^2\rho(\mathrm{d}\theta) \tag{1.7}$$

for $\psi \in \mathscr{C}$. Using the argument for deriving (1.6), we deduce from (1.7) that

$$\mathbb{E}|X(u; s_1, \xi)|^2 \lesssim 1 + \|\xi\|_\infty^2, \quad u \geq s_1.$$

This, together with (1.6), yields that

$$\mathbb{E}|\Gamma(t)|^2 \lesssim (1 + \|\xi\|_\infty^2)e^{-\lambda(t-s_2)}. \tag{1.8}$$

In what follows, we claim that

$$\mathbb{E}\|\Gamma_t\|_\infty^2 \lesssim (1 + \|\xi\|_\infty^2)e^{-\lambda(t-s_2)}. \tag{1.9}$$

For $r \in [t-\tau, t]$, let

$$\Lambda(t, r) = 2\int_{t-\tau}^{r} \langle \Gamma(u), \{\sigma(X_u(s_1, \xi)) - \sigma(X_u(s_2, \xi))\}\mathrm{d}\overline{W}(u)\rangle.$$

Applying the Burkholder–Davis–Gundy (B–D–G for abbreviation) inequality, taking (H2) into account and using the elementary inequality: $2ab \leq \varepsilon a^2 + \varepsilon^{-1}b^2$ for $a, b > 0$ and $\varepsilon > 0$ gives that

$$\begin{aligned}\mathbb{E}\|\Lambda_t\|_\infty &\lesssim \mathbb{E}\Big(\int_{t-\tau}^{t} |\Gamma(u)|^2 \cdot \|\sigma(X_u(s_1, \xi)) - \sigma(X_u(s_2, \xi))\|_{HS}^2\mathrm{d}u\Big)^{1/2}\\ &\leq \frac{1}{2}\mathbb{E}\|\Gamma_t\|_\infty^2 + c\int_{t-\tau}^{t} \mathbb{E}|\Gamma(u)|^2\mathrm{d}u,\end{aligned} \tag{1.10}$$

where $\|\Lambda_t\|_\infty := \sup_{-\tau\leq\theta\leq 0} |\Lambda(t, t+\theta)|$. Also, in the light of the Itô formula, and (H1), for any $r \in [t-\tau, t]$ we get that

$$|\Gamma(r)|^2 + (\lambda_1 - \lambda_2)\int_{t-\tau}^{r} |\Gamma(u)|^2\mathrm{d}u \leq |\Gamma(t-\tau)|^2 + \lambda_2\int_{t-2\tau}^{t-\tau} |\Gamma(u)|^2\mathrm{d}u + \Lambda(t, r).$$

Thus, we have proved that

$$\mathbb{E}\|\Gamma_t\|_\infty^2 \leq \mathbb{E}\|\Lambda_t\|_\infty + \mathbb{E}|\Gamma(t-\tau)|^2 + \lambda_2\int_{t-2\tau}^{t-\tau} \mathbb{E}|\Gamma(u)|^2\mathrm{d}u. \tag{1.11}$$

Consequently, (1.9) follows from (1.8), (1.10), and (1.11). Taking $s \to -\infty$, it follows that there exists an $\eta_t(\xi) \in L^2(\Omega \mapsto \mathscr{C})$ such that

$$\lim_{s\to-\infty} \mathbb{E}\|X_t(s, \xi) - \eta_t(\xi)\|_\infty^2 = 0. \tag{1.12}$$

Next, following the argument to derive (1.9), we obtain that

$$\mathbb{E}\|X_t(s,\xi)-X_t(s,\phi)\|_\infty^2 \lesssim \|\xi-\phi\|_\infty^2 e^{-\lambda(t-s)}. \tag{1.13}$$

Then, (1.12) and (1.13) yield that $\eta_t(\xi)$ is independent of the initial value $\xi \in \mathscr{C}$, which is denoted by η_t, by observing that

$$\begin{aligned}\mathbb{E}\|\eta_t(\xi)-\eta_t(\xi')\|_\infty \le& \mathbb{E}\|X_t(s,\xi)-\eta_t(\xi)\|_\infty + \mathbb{E}\|X_t(s,\xi')-\eta_t(\xi')\|_\infty \\ &+ \mathbb{E}\|X_t(s,\xi)-X_t(s,\xi')\|_\infty, \quad \xi,\xi' \in \mathscr{C}.\end{aligned}$$

For $-\infty < s \le t < \infty$, let

$$P_{s,t}(\xi,\cdot) = \mathbb{P}\circ(X_t(s,\xi))^{-1}(\cdot) \text{ and } P_{s,t}f(\xi) = \int_{\mathscr{C}} f(\psi)P_{s,t}(\xi,\mathrm{d}\psi), \ f \in \mathscr{B}_b(\mathscr{C}).$$

Then, for $-\infty < r \le s \le t < \infty$, by the Markov property of $X_t(s,\xi)$, we derive that

$$P_{r,s}\circ P_{s,t} = P_{r,t} \tag{1.14}$$

and, by the time homogeneity, that

$$P_{s,t}(\xi,\cdot) = P_{0,t-s}(\xi,\cdot). \tag{1.15}$$

By (1.12), it follows that $X_0(s,\xi)$ converges in probability to η_0 as $s \to -\infty$. Let $\phi \in C_b(\mathscr{C})$, the collection of all bounded continuous functions $f : \mathscr{C} \mapsto \mathbb{R}$. Then, we deduce from [74, Theorem 17.5, p. 145] that $\phi(X_0(s,\xi))$ converges in probability to $\phi(\eta_0)$ whenever $s \to -\infty$. So the dominated convergence theorem gives that $\mathbb{E}\phi(X_0(s,\xi)) \to \mathbb{E}\phi(\eta_0)$. Hence,

$$P_{-s,0}(\xi,\cdot) \to \mu := \mathbb{P}\circ\eta_0^{-1} \quad \text{weakly as } s \to \infty. \tag{1.16}$$

Let $f \in \mathscr{B}_b(\mathscr{C})$. Adopting the monotone class argument, we may assume $f \in L_b(\mathscr{C})$, the set of all bounded Lipschitz functions $f : \mathscr{C} \mapsto \mathbb{R}$. Because of $f \in L_b(\mathscr{C})$, in addition to (1.13), $P_{0,t}f \in C_b(\mathscr{C})$. Then, (1.14)–(1.16) and the definition of weak convergence of probability measures lead to

$$\mu(P_{0,t}f) = \lim_{s\to\infty} P_{-s,0}(P_{0,t}f)(\xi) = \lim_{s\to\infty} P_{-(t+s),0}f(\xi) = \mu(f). \tag{1.17}$$

Moreover, observe that $P_tf = P_{0,t}f$ for $t \ge 0$. Thus, (1.17) yields that $\mu = \mathbb{P}\circ\eta_0^{-1}$ is an invariant measure of (1.1).

Step 2: Uniqueness of invariant measure. Note that $\int_{\mathscr{C}} \|\xi'\|_\infty^2 \mu(\mathrm{d}\xi') < \infty$ due to $\eta_0 \in L^2(\Omega \mapsto \mathscr{C})$. Let $\mu' \in \mathscr{P}(\mathscr{C})$ be another invariant measure. Then,

$$\begin{aligned}|\mu(f)-\mu'(f)| &\leq \int_{\mathscr{C}}\int_{\mathscr{C}} |P_t f(\xi)-P_t f(\xi')|\mu(\mathrm{d}\xi)\mu'(\mathrm{d}\xi'), \quad f\in L_b(\mathscr{C})\\ &\lesssim e^{-\frac{\lambda t}{2}}\int_{\mathscr{C}}\int_{\mathscr{C}} \|\xi-\xi'\|_\infty \mu(\mathrm{d}\xi)\mu'(\mathrm{d}\xi')\\ &\to 0 \quad \text{as } t\to\infty.\end{aligned}$$

Thus, the uniqueness of invariant measure follows from [71, Proposition 2.2, p. 3].
Step 3: Exponential Mixing. By the invariance of μ and (1.13), we obtain that

$$\begin{aligned}|P_t f(\xi)-\mu(f)| &\leq \int_{\mathscr{C}} |P_t f(\xi)-P_t f(\xi')|\mu(\mathrm{d}\xi')\\ &\lesssim \int_{\mathscr{C}} (\mathbb{E}\|X_t(\xi)-X_t(\xi')\|_\infty^2)^{1/2}\mu(\mathrm{d}\xi')\\ &\lesssim e^{-\frac{\lambda t}{2}}\int_{\mathscr{C}} \|\xi-\xi'\|_\infty \mu(\mathrm{d}\xi'), \qquad f\in L(\mathscr{C}).\end{aligned}$$

As a result, the exponential mixing holds due to $\int_{\mathscr{C}} \|\xi'\|_\infty^2 \mu(\mathrm{d}\xi') < \infty$.

Remark 1.6 Under (H1) and (H2), by Theorem 1.5, the semigroup $(P_t)_{t\geq 0}$ has a unique invariant measure $\mu \in \mathscr{P}(\mathscr{C})$. Then, according to [37, Theorem 3.2.6, p. 28], the invariant measure μ is ergodic. Moreover, from [37, Theorem 3.2.4, p. 25], we derive that for any $f \in L^2(\mathscr{C},\mu)$, if $P_t f = f$, μ-a.s. for all $t \geq 0$, then f is a constant μ-a.s.; For any $\Gamma \in \mathscr{B}(\mathscr{C})$, if $P_t \mathbf{1}_\Gamma = \mathbf{1}_\Gamma$, μ-a.s. for all $t \geq 0$, then $\mu(\Gamma)=0$ or $\mu(\Gamma)=1$; For any $f \in L^2(\mathscr{C},\mu)$, $\frac{1}{T}\int_0^T P_t f \mathrm{d}t \to \mu(f)$ when $T\to\infty$.

Remark 1.7 The remote start method does not work well for an FSDE with variable delay

$$\mathrm{d}X(t) = b(X(t), X(t-r(t)))\mathrm{d}t + \sigma(X(t), X(t-r(t)))\mathrm{d}W(t), \quad t\geq 0$$

with the initial data $X_0=\xi\in\mathscr{C}$ since the functional solution $(X_t)_{t\geq 0}$ is an *inhomogeneous* Markov process.

1.4 Ergodicity for FSDEs of Neutral Type

Roughly speaking, an SDE is called an FSDE of neutral type if not only does it depend on the past and the present values but also involves derivatives with delays; see, e.g., [99, Chapter 6]. This kind of equation has also been utilized to model some evolution phenomena arising in, e.g., physics, biology, and engineering; see, e.g., Kolmanovskii–Nosov [81] for the theory in aeroelasticity, Mao [99] for the collision problem in electrodynamics, and Slemrod [131] for the oscillatory systems. Generally, an FSDE of neutral type admits the following form

$$\mathrm{d}\{X(t)-G(X_t)\}=b(X_t)\mathrm{d}t+\sigma(X_t)\mathrm{d}W(t),\quad t>0,\quad X_0=\xi\in\mathscr{C}. \tag{1.18}$$

Here $G:\mathscr{C}\to\mathbb{R}^n$ is refer to as a neutral term. In (1.18), b,σ,W are defined as in (1.1), and $G:\mathscr{C}\to\mathbb{R}^n$ is measurable, and locally continuous. Let $\rho(\cdot)$ be a probability measure on $[-\tau,0]$. In what follows, in addition to (H2), for any $\phi,\psi\in\mathscr{C}$, we further assume that the following assumptions hold.

(H3) There exists a $\kappa\in(0,1)$ such that

$$|G(\phi)-G(\psi)|^2\le\kappa\int_{[-\tau,0]}|\phi(\theta)-\psi(\theta)|^2\rho(\mathrm{d}\theta).$$

(H4) There exist $\nu_1>\nu_2>0$ such that

$$\begin{aligned}2\langle\phi(0)-\psi(0)-\{G(\phi)-G(\psi)\},b(\phi)-b(\psi)\rangle+\|\sigma(\phi)-\sigma(\psi)\|_{HS}^2\\ \le-\nu_1|\phi(0)-\psi(0)|^2+\nu_2\int_{[-\tau,0]}|\phi(\theta)-\psi(\theta)|^2\rho(\mathrm{d}\theta).\end{aligned}$$

Under conditions (H3) and (H4), (1.18) has a unique strong solution; see Theorem A.6.

Theorem 1.8 *Let* (H2)–(H4) *hold and assume further* $\kappa\in(0,1/4)$. *Then,* (1.18) *admits a unique invariant measure* μ, *which is exponentially mixing.*

Proof The proof of Theorem 1.8 is analogous to that of Theorem 1.5. However, for completeness, we provide a sketch of the proof below.

Let $(\overline{W}(t))_{t\in\mathbb{R}}$ be the double-sided Wiener process defined by (1.3). For fixed $s\in\mathbb{R}$, consider an FSDE of neutral type

$$\mathrm{d}\{X(t)-G(X_t)\}=b(X_t)\mathrm{d}t+\sigma(X_t)\mathrm{d}\overline{W}(t),\quad -\infty<s\le t<\infty \tag{1.19}$$

with the initial data $X_s=\xi\in\mathscr{C}$. Let $-\infty<s_1\le s_2\le t<\infty$, $\Gamma(t)$ be defined as in (1.5), and $N(t)=\Gamma(t)-\{G(X_t(s_1,\xi))-G(X_t(s_2,\xi))\}$. For any $\varepsilon>0$, by the Itô formula and (H4), we arrive at

$$\begin{aligned}e^{\varepsilon t}\mathbb{E}|N(t)|^2\le e^{\varepsilon s_2}\mathbb{E}|N(s_2)|^2+\varepsilon\int_{s_2}^t e^{\varepsilon r}\mathbb{E}|N(r)|^2\mathrm{d}r-\nu_1\int_{s_2}^t e^{\varepsilon r}\mathbb{E}|\Gamma(r)|^2\mathrm{d}r\\ +\nu_2\int_{s_2}^t e^{\varepsilon r}\int_{[-\tau,0]}\mathbb{E}|\Gamma(r+\theta)|^2\rho(\mathrm{d}\theta)\mathrm{d}r.\end{aligned} \tag{1.20}$$

According to the elementary inequality:

$$(a+b)^2\le(1+\alpha)(a^2+b^2/\alpha),\quad a,b,\alpha>0 \tag{1.21}$$

with $\alpha = \kappa$ below, it follows from (H3) that

$$\int_{s_2}^{t} e^{\varepsilon r}\mathbb{E}|N(r)|^2 dr \le (1+\kappa)\int_{s_2}^{t} e^{\varepsilon r}\Big\{\mathbb{E}|\Gamma(r)|^2 + \int_{[-\tau,0]} \mathbb{E}|\Gamma(r+\theta)|^2\rho(d\theta)\Big\}dr.$$

Plugging the previous estimate into (1.20) and taking (H3) into account yield that

$$\begin{aligned} e^{\varepsilon t}\mathbb{E}|N(t)|^2 \le\, & 2e^{\varepsilon s_2}\Big\{\mathbb{E}|\Gamma(s_2)|^2 + \kappa\int_{[-\tau,0]} \mathbb{E}|\Gamma(s_2+\theta)|^2\rho(d\theta)\Big\} \\ & + (\nu_2+\varepsilon(1+\kappa))e^{\varepsilon\tau}\int_{s_2-\tau}^{s_2} e^{\varepsilon r}\mathbb{E}|\Gamma(r)|^2 dr \\ & - \lambda_\varepsilon\int_{s_2}^{t} e^{\varepsilon r}\mathbb{E}|\Gamma(r)|^2 dr, \end{aligned} \tag{1.22}$$

where $\lambda_\varepsilon := \nu_1 - \varepsilon(1+\kappa) - (\nu_2 + \varepsilon(1+\kappa))e^{\varepsilon\tau}$. Next, following similar procedure to get (1.22), we can also show that

$$\sup_{t\ge s_1}\mathbb{E}|X(t; s_1,\xi)|^2 \lesssim 1 + \|\xi\|_\infty^2. \tag{1.23}$$

This further implies that

$$e^{\varepsilon t}\mathbb{E}|N(t)|^2 \le c(1+\|\xi\|_\infty^2)e^{\varepsilon s_2} - \lambda_\varepsilon\int_{s_2}^{t} e^{\varepsilon r}\mathbb{E}|\Gamma(r)|^2 dr. \tag{1.24}$$

Since $\nu_1 > \nu_2 > 0$ and $\kappa \in (0,1)$, we can choose $\varepsilon \in (0,1)$ sufficiently small and $\delta > 0$ sufficiently large such that

$$\lambda_\varepsilon > 0 \quad \text{and} \quad (1+\delta)\delta^{-1}\kappa\varepsilon^{\varepsilon\tau} < 1. \tag{1.25}$$

Hence, we deduce from (1.24) that

$$e^{\varepsilon t}\mathbb{E}|N(t)|^2 \lesssim e^{\varepsilon s_2}(1+\|\xi\|_\infty^2). \tag{1.26}$$

Moreover, for $\delta > 0$ such that (1.25), thanks to (1.21) and (H3), we deduce that

$$e^{\varepsilon t}\mathbb{E}|\Gamma(t)|^2 \le (1+\delta)e^{\varepsilon t}\mathbb{E}|N(t)|^2 + \frac{(1+\delta)\kappa}{\delta}e^{\varepsilon t}\int_{[-\tau,0]} \mathbb{E}|\Gamma(t+\theta)|^2\rho(d\theta).$$

Thus, for any $T > 0$, we obtain that

$$\begin{aligned}\sup_{s_2\le t\le T}(e^{\varepsilon t}\mathbb{E}|\Gamma(t)|^2) \le (1+\delta)&\sup_{s_2\le t\le T}(e^{\varepsilon t}\mathbb{E}|N(t)|^2)\\ &+(1+\delta)\delta^{-1}\kappa e^{\varepsilon\tau}\sup_{s_2-\tau\le t\le s_2}(e^{\varepsilon t}\mathbb{E}|\Gamma(t)|^2)\\ &+(1+\delta)\delta^{-1}\kappa e^{\varepsilon\tau}\sup_{s_2\le t\le T}(e^{\varepsilon t}\mathbb{E}|\Gamma(t)|^2).\end{aligned} \tag{1.27}$$

This, together with (1.23), (1.25), and (1.26), leads to

$$\begin{aligned}\sup_{s_2\le t\le T}(e^{\varepsilon t}\mathbb{E}|\Gamma(t)|^2) &\le \frac{1+\delta}{1-(1+\delta)\delta^{-1}\kappa e^{\varepsilon\tau}}\sup_{s_2\le t\le T}(e^{\varepsilon t}\mathbb{E}|N(t)|^2)\\ &\quad+\frac{(1+\delta)\delta^{-1}\kappa e^{\varepsilon\tau}}{1-(1+\delta)\delta^{-1}\kappa e^{\varepsilon\tau}}\sup_{s_2-\tau\le t\le s_2}(e^{\varepsilon t}\mathbb{E}|\Gamma(t)|^2)\\ &\lesssim e^{\varepsilon s_2}(1+\|\xi\|_\infty^2).\end{aligned}$$

Taking $T\to\infty$ gives that

$$\mathbb{E}|\Gamma(t)|^2 \lesssim e^{-\varepsilon(t-s_2)}(1+\|\xi\|_\infty^2). \tag{1.28}$$

In view of the Itô formula, the B–D–G inequality, (1.26) and (1.28), we obtain that

$$\mathbb{E}\|N_t\|_\infty^2 \lesssim e^{-\varepsilon(t-s_2)}(1+\|\xi\|_\infty^2). \tag{1.29}$$

In what follows, without loss of generality, we may assume that $t = n\tau$, $s_1=-(m+1)\tau$ and $s_2=-m\tau$, where n and m are positive integers. By (1.21) with $\delta=(1-\sqrt{\kappa})/\sqrt{\kappa}$ and (H3), it follows that

$$\begin{aligned}&\mathbb{E}\|X_{n\tau}(-m\tau,\xi)-X_{n\tau}(-(m+1)\tau,\xi)\|_\infty^2\\ &\le\sqrt{\kappa}\mathbb{E}\|X_{n\tau}(-m\tau,\xi)-X_{n\tau}(-(m+1)\tau,\xi)\|_\infty^2\\ &\quad+\sqrt{\kappa}\mathbb{E}\|X_{(n-1)\tau}(-m\tau,\xi)-X_{(n-1)\tau}(-(m+1)\tau,\xi)\|_\infty^2+\frac{ce^{-\varepsilon(n+m)\tau}(1+\|\xi\|_\infty^2)}{1-\sqrt{\kappa}}.\end{aligned}$$

Thus, by an induction argument we conclude that

$$\mathbb{E}\|X_{n\tau}(-m\tau,\xi)-X_{n\tau}(-(m+1)\tau,\xi)\|_\infty^2\lesssim\left\{\Big(\frac{\sqrt{k}}{1-k}\Big)^{n+m}+\frac{e^{\epsilon(n+m)\tau}}{1-q}\right\}(1+\|\xi\|_\infty^2)$$

in which $q:=\sqrt{\kappa}e^{\varepsilon\tau}/(1-\sqrt{\kappa})<1$ for $\kappa\in(0,1/4)$ and sufficiently small $\varepsilon\in(0,1)$. As a result, there exists $\gamma>0$ such that

$$\mathbb{E}\|X_{n\tau}(-m\tau,\xi)-X_{n\tau}(-(m+1)\tau,\xi)\|_\infty^2\lesssim e^{-\gamma(n+m)\tau}(1+\|\xi\|_\infty^2).$$

The remainder of the proof follows from the lines of the argument for Theorem 1.5.

1.5 Ergodicity for FSPDEs Driven by Cylindrical Wiener Processes

Let $(H, \langle \cdot, \cdot \rangle_H, \| \cdot \|_H)$ be a real separable Hilbert space, and $(W(t))_{t \geq 0}$ a cylindrical Wiener process on H under a complete filtered probability space $(\Omega, \mathscr{F}, (\mathscr{F}_t)_{t \geq 0}, \mathbb{P})$ so that

$$W(t) = \sum_{k=1}^{\infty} B_k(t) e_k, \quad t \geq 0$$

for an orthonormal basis $(e_k)_{k \geq 1}$ on H and a sequence of i.i.d. (independent and identically distributed) $\mathbb{R}$-valued Wiener processes $(B_k(t))_{k \geq 1}$ on $(\Omega, \mathscr{F}, (\mathscr{F}_t)_{t \geq 0}, \mathbb{P})$. Let $\mathscr{L}(H)$ and $\mathscr{L}_{HS}(H)$ be the spaces of all bounded linear operators and Hilbert–Schmidt operators on H, respectively. Denote by $\| \cdot \|$ and $\| \cdot \|_{HS}$ the operator norm and the Hilbert–Schmidt norm, respectively. Let $\mathscr{C} = C([-\tau, 0]; H)$, the space of all continuous functions $f : [-\tau, 0] \mapsto H$, equipped with the uniform norm $\|f\|_\infty := \sup_{-\tau \leq \theta \leq 0} \|f(\theta)\|_H$. Let $\rho(\cdot)$ be a probability measure on $[-\tau, 0]$.

Consider a semi-linear FSPDE on $(H, \langle \cdot, \cdot \rangle_H, \| \cdot \|_H)$ in the form

$$\mathrm{d}X(t) = \{AX(t) + b(X_t)\}\mathrm{d}t + \sigma(X_t)\mathrm{d}W(t), \quad X_0 = \xi \in \mathscr{C}. \tag{1.30}$$

Herein, $b : \mathscr{C} \mapsto H$ and $\sigma(\zeta) := \sigma_0 + \sigma_1(\zeta), \zeta \in \mathscr{C}$, where $\sigma_0 \in \mathscr{L}(H)$ and $\sigma_1 : \mathscr{C} \mapsto \mathscr{L}_{HS}(H)$. We assume the following conditions.

(A1) $(-A, \mathscr{D}(A))$ is self-adjoint with discrete spectrum $0 < \lambda_1 \leq \lambda_2 \leq \cdots \leq \lambda_i \leq \cdots$ counting multiplicities such that $\lambda_i \uparrow \infty$. In this case, A generates a C_0 semigroup $(e^{tA})_{t \geq 0}$ such that $\|e^{tA}\| \leq e^{-\lambda_1 t}, t \geq 0$.

(A2) There exists an $L > 0$ such that, for any $\phi, \psi \in \mathscr{C}$,

$$\begin{aligned}
&\|b(\phi) - b(\psi)\|_H^2 + \|\sigma_1(\phi) - \sigma_1(\psi)\|_{HS}^2 \\
&\leq L\Big(\|\phi(0) - \psi(0)\|_H^2 + \int_{[-\tau,0]} \|\phi(\theta) - \psi(\theta)\|_H^2 \rho(\mathrm{d}\theta)\Big).
\end{aligned}$$

(A3) There exist constants $\nu_1 \in \mathbb{R}$ and $\nu_2 > 0$ such that

$$\begin{aligned}
&2\langle \phi(0) - \psi(0), b(\phi) - b(\psi) \rangle_H + \|\sigma_1(\phi) - \sigma_1(\psi)\|_{HS}^2 \\
&\leq \nu_1 \|\phi(0) - \psi(0)\|_H^2 + \nu_2 \int_{[-\tau,0]} \|\phi(\theta) - \psi(\theta)\|_H^2 \rho(\mathrm{d}\theta), \qquad \phi, \psi \in \mathscr{C}.
\end{aligned}$$

(A4) $\sup_{t \geq 0} \Big(\int_0^t \|e^{sA}\sigma_0\|_{HS}^2 \mathrm{d}s \Big) < \infty$.

Remark 1.9 For a bounded domain $D \subset \mathbb{R}^d$, let $H = L^2(D; \mathrm{d}x)$ and $A = -(-\Delta)^\alpha$, where Δ is the Laplacian on D and $\alpha > \frac{d}{2}$ is a constant. Let $\sigma = I$ be the identity operator on H, and $b(\xi) = L \int_{[-\tau,0]} \xi(r)\nu(\mathrm{d}r)$ for a signed measure ν on $[-\tau, 0]$

with the total variation 1. Since $A = -(-\Delta)^{\alpha}$, it is well known that the eigenvalues $(\lambda_i)_{i\geq 1}$ of A satisfy $\lambda_i \geq c i^{\frac{2\alpha}{d}}$ $(i \geq 1)$ for some constant $c > 0$. Then, (A1) holds.

By Theorem A.9, (1.30) admits a unique mild solution (see, also e.g., [11, Theorem A.1]), i.e., there exists a unique continuous adapted process $(X(t))_{t\geq -\tau}$ on H such that $\mathbb{P}$-a.s.

$$X(t) = e^{tA}\xi(0) + \int_0^t e^{(t-s)A} b(X_s)\mathrm{d}s + \int_0^t e^{(t-s)A}\sigma(X_s)\mathrm{d}W(s), \quad t > 0.$$

Theorem 1.10 *Let* (A1)–(A4) *hold and assume further* $2\lambda_1 - \nu_1 > \nu_2$. *Then,* (1.30) *has a unique invariant measure* $\mu \in \mathscr{P}(\mathscr{C})$*, which is exponentially mixing.*

Proof Let $\widetilde{W}(t) = \sum_{k=1}^{\infty} \beta_k \widetilde{B}_k(t) e_k$, where $(\widetilde{B}_k(t))_{k\geq 0}$ is an independent copy of $(B_k(t))_{t\geq 0}$, and $(\overline{W}(t))_{t\geq 0}$ be a double-sided cylindrical Wiener process defined by

$$\overline{W}(t) = \begin{cases} W(t), & t \geq 0, \\ \widetilde{W}(-t), & t < 0 \end{cases}$$

with the filtration

$$\overline{\mathscr{F}}_t := \bigcap_{s>t} \bar{\mathscr{F}}_s^0,$$

where $\overline{\mathscr{F}}_s^0 := \sigma(\{\overline{W}(r_2) - \overline{W}(r_1) : -\infty < r_1 \leq r_2 \leq s\}, \mathscr{N})$ and $\mathscr{N} := \{A \in \mathscr{F} | \mathbb{P}(A) = 0\}$. Fix $s \in \mathbb{R}$ and consider the following semi-linear FSPDE

$$\mathrm{d}X(t) = \{AX(t) + b(X_t)\}\mathrm{d}t + \sigma(X_t)\mathrm{d}\overline{W}(t), \quad t > s, \quad X_s = \xi. \tag{1.31}$$

Following the line of proving Theorem 1.5, to complete the proof, it is sufficient to verify that

$$\mathbb{E}\|\Gamma_t\|_{\infty}^2 \lesssim (1 + \|\xi\|_{\infty}^2)e^{-\lambda(t-s_2)}, \tag{1.32}$$

where $\Gamma(t)$ is defined as in (1.5). Set

$$Z(t,s) := \int_s^t e^{(t-r)A}\sigma_0 \mathrm{d}\overline{W}(r), \; t \geq s, \; Z(t,s) :\equiv 0, \; t < s, \; Y(t,s) := X(t;s,\xi) - Z(t,s).$$

Thus, (1.31) can be reformulated as

$$\mathrm{d}Y(t,s) = \{AY(t,s) + b(Y_t(s) + Z_t(s))\}\mathrm{d}t + \sigma_1(Y_t(s) + Z_t(s))\mathrm{d}\overline{W}(t), \; t > s, Y_s = \xi \in \mathscr{C}.$$

To proceed, we aim to show that

$$\sup_{t\geq s} \mathbb{E}\|Y(t,s)\|_H^2 \lesssim 1 + \|\xi\|_{\infty}^2. \tag{1.33}$$

For any $\varepsilon > 0$, it follows from (A2) and (A3) that there exists $c_\varepsilon > 0$ such that

$$2\langle \xi(0), b(\xi)\rangle_H + \|\sigma_1(\xi)\|_{HS}^2 \le c_\varepsilon + (\nu_1+\varepsilon)\|\xi(0)\|_H^2 + (\nu_2+\varepsilon)\int_{[-\tau,0]} \|\xi(\theta)\|_H^2 \rho(\mathrm{d}\theta) \tag{1.34}$$

for $\xi, \eta \in \mathscr{C}$. According to [92, Proposition 2.1.4, p. 51], we obtain from (A1) that

$$\langle Ax, x\rangle_H \le -\lambda_1 \|x\|_H^2, \quad x \in \mathscr{D}(A). \tag{1.35}$$

Employing the Itô formula, and taking (A2), (1.34), and (1.35) into consideration, for any $\lambda \in (0, 2\lambda_1)$ we derive that

$$\begin{aligned}
&e^{\lambda t}\mathbb{E}\|Y(t,s)\|_H^2\\
&\le e^{\lambda s}\|\xi(0)\|_H^2 + \int_s^t e^{\lambda r}\mathbb{E}\{(\lambda - 2\lambda_1)\|Y(r,s)\|_H^2\\
&+ 2\langle Y(r,s), b(Y_r(s)+Z_r(s))\rangle_H + \|\sigma_1(Y_r(s)+Z_r(s))\|_{HS}^2\}\mathrm{d}r\\
&\le e^{\lambda s}\|\xi(0)\|_H^2 + \int_s^t e^{\lambda r}\mathbb{E}\Big\{c_\varepsilon + (\lambda-2\lambda_1)\|Y(r,s)\|_H^2 + (\nu_1+\varepsilon)\|Y(r,s)+Z(r,s)\|_H^2\\
&+ (\nu_2+\varepsilon)\int_{[-\tau,0]} \|Y(r+\theta,s)+Z(r+\theta,s)\|_H^2\rho(\mathrm{d}\theta)\\
&+ 2\|Z(r,s)\|_H\cdot\|b(Y_r(s)+Z_r(s))\|_H\Big\}\mathrm{d}r\\
&\le \frac{c_\varepsilon}{\lambda}(e^{\lambda t}-e^{\lambda s}) + e^{\lambda s}\|\xi(0)\|_H^2 + (\nu_2+2\varepsilon)e^{\lambda\tau}\int_{s-\tau}^s e^{\lambda r}\|\xi(r)\|_H^2\mathrm{d}s\\
&+ \frac{L+(2\lambda_1-\lambda)(1-\varepsilon)}{\varepsilon}\mathbb{E}\int_s^t e^{\lambda s}\|Z(r,s)\|_H^2\mathrm{d}r\\
&- \{(2\lambda_1-\lambda)(1-\varepsilon)-(\nu_1+2\varepsilon)-(\nu_2+2\varepsilon)e^{\lambda\tau}\}\int_s^t e^{\lambda r}\mathbb{E}\|Y(r,s)+Z(r,s)\|_H^2\mathrm{d}r.
\end{aligned}$$

Because $2\lambda_1 - \nu_1 > \nu_2$, we choose $\varepsilon > 0$ and $\lambda \in (0, 2\lambda_1)$ sufficiently small such that

$$(2\lambda_1-\lambda)(1-\varepsilon)-(\nu_1+2\varepsilon)-(\nu_2+2\varepsilon)e^{\lambda\tau} = 0$$

so that

$$e^{\lambda t}\mathbb{E}\|Y(t,s)\|_H^2 \lesssim \frac{c_\varepsilon}{\lambda}(e^{\lambda t}-e^{\lambda s}) + e^{\lambda s}\|\xi\|_\infty^2 + \int_s^t e^{\lambda r}\mathbb{E}\|Z(r,s)\|_H^2\mathrm{d}r. \tag{1.36}$$

Next, by the Itô isometry, in addition to (A4), one finds that

$$\sup_{t\geq s}\mathbb{E}\|Z(t,s)\|_H^2=\sup_{t\geq s}\Big(\int_s^t\|e^{(t-u)A}\sigma_0\|_{HS}^2\mathrm{d}u\Big)$$
$$=\sup_{t\geq s}\Big(\int_0^{t-s}\|e^{(t-s-u)A}\sigma_0\|_{HS}^2\mathrm{d}u\Big)<\infty. \tag{1.37}$$

Thus, (1.33) follows from (1.36) and (1.37) immediately.

For any $\lambda\in(0,2\lambda_1)$, performing Itô's formula and utilizing (1.35) and (A3) yield that

$$\begin{aligned}e^{\lambda t}\mathbb{E}\|\Gamma(t)\|_H^2&=e^{\lambda s_2}\mathbb{E}\|\Gamma(s_2)\|_H^2+\int_{s_2}^t e^{\lambda r}\mathbb{E}\{\lambda\|\Gamma(r)\|_H^2\\&\quad+2\langle\Gamma(r),A\Gamma(r)+b(X_r(s_1,\xi))-b(X_r(s_2,\xi))\rangle_H\\&\quad+\|\sigma_1(X_r(s_1,\xi))-\sigma_1(X_r(s_2,\xi))\|_{HS}^2\}\mathrm{d}r\\&\leq e^{\lambda s_2}\mathbb{E}\|\Gamma(s_2)\|_H^2+\int_{s_2}^t e^{\lambda r}\mathbb{E}\Big\{(\lambda-2\lambda_1+\nu_1)\|\Gamma(r)\|_H^2\\&\quad+\nu_2\int_{[-\tau,0]}\|\Gamma(r+\theta)\|_H^2\rho(\mathrm{d}\theta)\Big\}\mathrm{d}r\\&\leq e^{\lambda s_2}\mathbb{E}\|\Gamma(s_2)\|_H^2+\nu_2e^{\lambda\tau}\int_{s_2-\tau}^{s_2}e^{\lambda r}\mathbb{E}\|\Gamma(r)\|_H^2\mathrm{d}r\\&\quad-(2\lambda_1-\nu_1-\lambda-\nu_2e^{\lambda\tau})\int_{s_2}^t e^{\lambda r}\mathbb{E}\|\Gamma(r)\|_H^2\mathrm{d}r.\end{aligned}$$

Taking $\lambda=2\lambda_1-\nu_1-\nu_2e^{\lambda\tau}>0$ and using $2\lambda_1-\nu_1>\nu_2$ yield that

$$e^{\lambda t}\mathbb{E}\|\Gamma(t)\|_H^2\leq e^{\lambda s_2}\mathbb{E}\|\Gamma(s_2)\|_H^2+\nu_2e^{\lambda\tau}\int_{s_2-\tau}^{s_2}e^{\lambda r}\mathbb{E}\|\Gamma(r)\|_H^2\mathrm{d}r.$$

This, together with (1.33) and (1.37), leads to

$$\mathbb{E}\|\Gamma(t)\|_H^2\lesssim(1+\|\xi\|_\infty^2)e^{-\lambda(t-s_2)}. \tag{1.38}$$

Next, (A1) and [36, Theorem 6.10, p. 160] lead to

$$\begin{aligned}\mathbb{E}\|\Gamma_t\|_\infty^2&\lesssim\mathbb{E}\|\Gamma(t-\tau)\|_H^2+\int_{t-\tau}^t\mathbb{E}\|b(X_u(s_1,\xi))-b(X_u(s_2,\xi))\|_H^2\mathrm{d}u\\&\quad+\mathbb{E}\Big(\sup_{t-\tau\leq r\leq t}\Big\|\int_{t-\tau}^r e^{(r-u)A}\{\sigma_1(X_u(s_1,\xi))-\sigma_1(X_u(s_2,\xi))\}\mathrm{d}\overline{W}(u)\Big\|_H^2\Big)\\&\lesssim\mathbb{E}\|\Gamma(t-\tau)\|_H^2+\int_{t-\tau}^t\mathbb{E}\{\|b(X_u(s_1,\xi))-b(X_u(s_2,\xi))\|_H^2\end{aligned}$$

$$
\begin{aligned}
&+\sigma_1(X_u(s_1,\xi))-\sigma_1(X_u(s_2,\xi))\|_{HS}^2\}\mathrm{d}u \\
&\lesssim \mathbb{E}\|\Gamma(t-\tau)\|_H^2+\int_{t-2\tau}^t \mathbb{E}\|\Gamma(r)\|_H^2\mathrm{d}r \\
&\lesssim (1+\|\xi\|_\infty^2)e^{-\lambda(t-s_2)},
\end{aligned}
$$

where we have used (A2) in the last two step, and utilized (1.38) in the last step. Consequently, (1.32) follows.

1.6 Ergodicity for FSDEs Driven by Jump Processes

Although there are a great number of references on existence of invariant measures for functional stochastic equations driven by (cylindrical) Brownian motions, the results on functional stochastic equations driven by jump processes are relatively scarce. Reiß et al. [119] explored existence of invariant measures for a class of semi-linear FSDEs driven by jump processes by considering the semi-martingale characteristics, where the jump diffusion term is uniformly bounded. Taking advantage of some exponential-type estimates, Bo-Yuan [21] investigated existence and uniqueness of invariant measures for stochastic differential delay equations (SDDEs) with jumps. However, their techniques are dimension dependent and hence difficult to be adopted to infinite-dimensional FSPDEs driven by jump processes. For stochastic systems with continuous sample paths, stemming from the Arzelà–Ascoli theorem, Kolmogorov's tightness criterion [77, Problem 2.4.11, p. 64] plays a crucial role. In contrast, for diffusion processes with jumps, such tightness criterion is not applicable.

In the previous sections, we concentrated on ergodicity for functional stochastic equations driven by (cylindrical) Brownian motions. In this section, we proceed to investigate ergodicity for FSDEs driven by jump processes. To begin with, we introduce some notation. Let $\mathscr{D}=D([-\tau,0];\mathbb{R}^n)$ denote the collection of all càdlàg paths $f:[-\tau,0]\mapsto\mathbb{R}^n$. Recall that a path $f:[-\tau,0]\mapsto\mathbb{R}^n$ is càdlàg if it is right continuous having finite left-hand limits. Let Λ denote the class of increasing homeomorphisms, and

$$
\|\lambda\|^\circ=\sup_{-\tau\le s<t\le 0}\left|\log\frac{\lambda(t)-\lambda(s)}{t-s}\right|<\infty.
$$

Under the uniform metric $\|\zeta\|_\infty=\sup_{-\tau\le\theta\le 0}|\zeta(\theta)|$ for each $\zeta\in\mathscr{D}$, the space $\mathscr{D}$ is complete but not separable. For $\xi,\eta\in\mathscr{D}$, define the Skorohod metric d_S on $\mathscr{D}$ by

$$
\mathrm{d}_S(\xi,\eta)=\inf_{\lambda\in\Lambda}\{\|\lambda\|^\circ\vee\|\xi-\eta\circ\lambda\|_\infty\}.
$$

Under the Skorohod metric d_S, $\mathscr{D}$ is complete and separable [20, Theorem 12.2, p. 128]. For the space $\mathscr{D}$, the uniform metric $\|\cdot\|_\infty$ is not as convenient as that of the Skorohod metric d_S. For more details on the Skorohod metric, we refer to [20, Chapter 4]. Let $(\mathbb{Y}, \mathscr{B}(\mathbb{Y}))$ be a measurable space, and $p : D_p \mapsto \mathbb{Y}$ an adapted process, where D_p is a countable subset of $\mathbb{R}_+$. Then, as in Ikeda–Watanabe [71, p. 59], the Poisson random measure $N(\cdot,\cdot) : \mathscr{B}(\mathbb{R}_+ \times \mathbb{Y}) \times \Omega \mapsto \mathbb{N} \cup \{0\}$, defined on the complete filtered probability space $(\Omega, \mathscr{F}, (\mathscr{F}_t)_{t\geq 0}, \mathbb{P})$, can be represented by

$$N((0,t] \times \Gamma) = \sum_{s \in D_p, s \leq t} \mathbf{1}_\Gamma(p(s)), \quad \Gamma \in \mathscr{B}(\mathbb{Y}).$$

In this case, we say that p is a Poisson point process and N is the Poisson random measure generated by p. Let $\nu(\cdot) = \mathbb{E}N((0,1] \times \cdot)$. Then, the compensated Poisson random measure

$$\widetilde{N}(\mathrm{d}t, \mathrm{d}z) := N(\mathrm{d}t, \mathrm{d}z) - \mathrm{d}t\,\nu(\mathrm{d}z) \quad \text{is a martingale.}$$

Consider an FSDE with jumps on $(\mathbb{R}^n, \langle\cdot,\cdot\rangle, |\cdot|)$ of the form

$$\mathrm{d}X(t) = b(X_t)\mathrm{d}t + \int_\Gamma \gamma(X_{t-}, z)\widetilde{N}(\mathrm{d}t, \mathrm{d}z), \quad t > 0, \quad X_0 = \xi \in \mathscr{D}, \tag{1.39}$$

where $b : \mathscr{D} \mapsto \mathbb{R}^n$ is locally Lipschitz, $\gamma : \mathscr{D} \times \Gamma \mapsto \mathbb{R}^n$ is locally Lipschitz with respect to the first variable, and $X_{t-}(\theta) := X((t+\theta)-) := \lim_{s \uparrow t+\theta} X(s)$ for $\theta \in [-\tau, 0]$. Moreover, we assume that the initial data $X_0 = \xi$ is independent of $N(\cdot,\cdot)$. Let $\rho(\cdot)$ be a probability measure on $[-\tau, 0]$. In this section, we assume that $\mathscr{D}$ is endowed with the Skorohod topology. For any $\phi, \psi \in \mathscr{D}$, we assume the following conditions hold.

(B1) There exist constants $\beta_1 > \beta_2 > 0$ such that

$$\begin{aligned}
&2\langle \phi(0) - \psi(0), b(\phi) - b(\psi)\rangle + \int_\Gamma |\gamma(\phi, z) - \gamma(\psi, z)|^2 \nu(\mathrm{d}z) \\
&\leq -\beta_1 |\phi(0) - \psi(0)|^2 + \beta_2 \int_{[-\tau,0]} |\phi(\theta) - \psi(\theta)|^2 \rho(\mathrm{d}\theta);
\end{aligned}$$

(B2) There exists a constant $\beta_3 > 0$ such that

$$\begin{aligned}
\int_\Gamma |\gamma(\phi, z) - \gamma(\psi, z)|^2 \nu(\mathrm{d}z) \leq \beta_3 \Big\{ &|\phi(0) - \psi(0)|^2 \\
&+ \int_{[-\tau,0]} |\phi(\theta) - \psi(\theta)|^2 \rho(\mathrm{d}\theta) \Big\}.
\end{aligned}$$

Under (B1) and (B2), there is a unique solution for (1.39); see Theorem A.8.

Theorem 1.11 *Let* (B1)–(B2) *hold and assume further*

$$\int_{\Gamma} |\gamma(0,z)|^2 \nu(dz) < \infty. \tag{1.40}$$

Then, (1.39) *admits a unique invariant measure* $\pi \in \mathscr{P}(\mathscr{D})$.

Proof Again, we adopt the remote start approach to explore existence of an invariant measure for (1.39). Let $N_1(\cdot,\cdot)$ be an independent copy of $N(\cdot,\cdot)$ and $N_0(\cdot,\cdot)$ a double-sided Poisson process defined by

$$N_0(t,\Gamma) = \begin{cases} N(t,\Gamma), & t \ge 0, \\ N_1(-t,\Gamma), & t < 0 \end{cases}$$

for all $\Gamma \in \mathscr{B}(\mathbb{Y})$ with the filtration

$$\overline{\mathscr{F}}_t := \bigcap_{s>t} \overline{\mathscr{F}}^0_s,$$

where $\overline{\mathscr{F}}^0_s := \sigma(\{N_0(r_2,\Gamma) - N_0(r_1,\Gamma) : -\infty < r_1 \le r_2 \le s, \Gamma\}, \mathscr{N})$ and $\mathscr{N} := \{A \in \mathscr{F} | \mathbb{P}(A) = 0\}$. For fixed $s \in \mathbb{R}$ and $\xi \in \mathscr{D}$, consider an FSDE

$$\mathrm{d}X(t) = b(X_t)\mathrm{d}t + \int_{\Gamma} \gamma(X_{t-},z)\widetilde{N}_0(\mathrm{d}t,\mathrm{d}z), \quad t > s, \quad X_s = \xi, \tag{1.41}$$

where $\widetilde{N}_0(\mathrm{d}t,\mathrm{d}z) := N_0(\mathrm{d}t,\mathrm{d}z) - \mathrm{d}t\nu(\mathrm{d}z)$. Under (B1)–(B2), (1.41) admits a unique strong solution $X(t;s,\xi)$ with the initial data $X_s = \xi$. If we can claim that

$$\mathbb{E}\|\Gamma_t\|_\infty^2 \lesssim \Big(1 + |\xi(0)|^2 + \int_{-\tau}^0 |\xi(\theta)|^2 \mathrm{d}\theta\Big) e^{-\nu(t-s_2)}, \tag{1.42}$$

where $\Gamma(t)$ is defined as in (1.5), together with

$$\mathrm{d}_S(\xi,\eta) \le \|\xi - \eta\|_\infty, \quad \xi,\eta \in \mathscr{D},$$

then we conclude that $(X_t(s,\xi))_{s<0}$ is a Cauchy sequence in $L^2(\Omega \mapsto \mathscr{D})$. Thus, (1.39) admits an invariant measure $\pi \in \mathscr{P}(\mathscr{D})$ by following the argument of that in Step 1 of Theorem 1.5. In what follows, it is sufficient to verify the estimate (1.42). By (B1)–(B2), besides (1.40), it is readily seen that there exist $\nu_1 > \nu_2 > 0$ such that

$$\begin{aligned} &2\langle \phi(0), b(\phi)\rangle + \int_{\Gamma} |\gamma(\phi,z)|^2 \nu(\mathrm{d}z) \\ &\le -\nu_1|\phi(0)|^2 + \nu_2 \int_{[-\tau,0]} |\phi(\theta)|^2 \rho(\mathrm{d}\theta) + c, \quad \phi \in \mathscr{D}. \end{aligned} \tag{1.43}$$

Carrying out the argument to get (1.6), we deduce from (1.43) that

$$\mathbb{E}|X(t;s,\xi)|^2 \lesssim 1+|\xi(0)|^2+\int_{-\tau}^{0}|\xi(\theta)|^2\mathrm{d}\theta,\ \ t\ge s, \tag{1.44}$$

and from (B1) that

$$\mathbb{E}|\Gamma(t)|^2 \lesssim \Big(1+|\xi(0)|^2+\int_{-\tau}^{0}|\xi(\theta)|^2\mathrm{d}\theta\Big)e^{-\nu(t-s_2)}, \tag{1.45}$$

Applying the Itô formula and utilizing (B1), we have

$$\mathbb{E}\|\Gamma_t\|_\infty^2 \le \mathbb{E}\|M_t\|_\infty + \mathbb{E}|\Gamma(t-\tau)|^2+\beta_2\int_{t-2\tau}^{t-\tau}\mathbb{E}|\Gamma(u)|^2\mathrm{d}u, \tag{1.46}$$

where, for any $r\in[t-\tau,t]$,

$$\begin{aligned}M(r):=\int_{t-\tau}^{r}\int_{\Gamma}\{&2\langle\Gamma(u-),\gamma(X_{u-}(s_1,\xi),z)-\gamma(X_{u-}(s_2,\xi),z)\rangle\\&+|\gamma(X_{u-}(s_1,\xi),z)-\gamma(X_{u-}(s_2,\xi),z)|^2\}\widetilde{N}_0(\mathrm{d}u,\mathrm{d}z)\end{aligned}$$

with $\Gamma(t-):=X(t-;s_1,\xi)-X(t-;s_2,\xi)$. Let $[M,M](r)$ be the quadratic variation process of $M(r)$. By the B–D–G inequality [116, Theorem 48, p. 193], and the definition of stochastic integral w.r.t. the Poisson jump process, we obtain that

$$\begin{aligned}\mathbb{E}\|M_t\|_\infty &\lesssim \mathbb{E}([M,M](t))^{1/2}\\&=\mathbb{E}\Big\{\int_{t-\tau}^{t}\int_{\Gamma}\{2\langle\Gamma(u-),\gamma(X_{u-}(s_1,\xi),z)-\gamma(X_{u-}(s_2,\xi),z)\rangle\\&\quad+|\gamma(X_{u-}(s_1,\xi),z)-\gamma(X_{u-}(s_2,\xi),z)|^2\}^2N_0(\mathrm{d}u,\mathrm{d}z)\Big\}^{1/2}\\&=\mathbb{E}\Big\{\sum_{t-\tau\le u\le t,u\in D_p,p(u)\in\Gamma}\{2<\Gamma(u),\gamma(X_u(s_1,\xi),p(u))-\gamma(X_u(s_2,\xi),p(u))\rangle\\&\quad+|\gamma(X_u(s_1,\xi),p(u))-\gamma(X_u(s_2,\xi),p(u))|^2\}^2\Big\}^{1/2}.\end{aligned}$$

Thus, a straightforward calculation shows that

$$\begin{aligned}\mathbb{E}\|M_t\|_\infty \lesssim\ &\mathbb{E}\Big\{\|\Gamma_t\|_\infty^2\sum_{t-\tau\le u\le t,u\in D_p,p(u)\in\Gamma}|\gamma(X_u(s_1,\xi),p(u))-\gamma(X_u(s_2,\xi),p(u))|^2\Big\}^{\frac{1}{2}}\\&+\mathbb{E}\Big(\sum_{t-\tau\le u\le t,u\in D_p,p(u)\in\Gamma}|\gamma(X_u(s_1,\xi),p(u))-\gamma(X_u(s_2,\xi),p(u))|^2\Big).\end{aligned}$$

Consequently, we obtain from (B2) that

$$\begin{aligned}
&\mathbb{E}\|M_t\|_\infty \\
&\le \frac{1}{2}\|\Gamma_t\|_\infty^2 + c\mathbb{E}\Big(\int_{t-\tau}^t \int_\Gamma |\gamma(X_{r-}(s_1,\xi),z) - \gamma(X_{r-}(s_2,\xi),z)|^2 N_0(\mathrm{d}r,\mathrm{d}z)\Big) \\
&\le \frac{1}{2}\|\Gamma_t\|_\infty^2 + c\int_{t-\tau}^t \int_\Gamma \mathbb{E}|\gamma(X_r(s_1,\xi),z) - \gamma(X_r(s_2,\xi),z)|^2 \nu(\mathrm{d}z)\mathrm{d}r \qquad (1.47) \\
&\le \frac{1}{2}\|\Gamma_t\|_\infty^2 + c\int_{t-2\tau}^t \mathbb{E}|\Gamma(r)|^2 \mathrm{d}r
\end{aligned}$$

Thus, (1.42) holds by substituting (1.47) into (1.46) and taking (1.45) into consideration. Moreover, following the argument to deduce (1.42), we have

$$\mathbb{E}\|X_t(\xi) - X_t(\eta)\|_\infty^2 \lesssim \Big(|\xi(0) - \eta(0)|^2 + \int_{-\tau}^0 |\xi(\theta) - \eta(\theta)|^2 \mathrm{d}\theta\Big) e^{-\lambda t}$$

for some $\lambda > 0$. As a result, the uniqueness of invariant measure follows.

1.7 Ergodicity for FSPDEs Driven by α-Stable Processes

Here we focus on ergodicity of FSPDEs driven by cylindrical α-stable processes. We further need to introduce some notions and notation. Let $(H, \langle\cdot,\cdot\rangle, \|\cdot\|_H)$ be a real separable Hilbert space. Recall that a random variable η is said to be stable with stability index α, scale parameter $\sigma \in (0,\infty)$, skewness parameter $\beta \in [-1,1]$, and location parameter $\mu \in (-\infty,\infty)$ if it has characteristic function of the form

$$\begin{aligned}
\phi_\eta(u) &= \mathbb{E}\exp(iu\eta) \\
&= \begin{cases} \exp\{-\sigma^\alpha|u|^\alpha(1 - i\beta(\mathrm{sgn}(u))\tan(\pi\alpha/2)) + iu\mu\} & \text{if } \alpha \neq 1, \\ \exp\{-\sigma^\alpha|u|^\alpha(1 + i\beta\frac{2}{\pi}(\mathrm{sgn}(u))\log|u|) + iu\mu\} & \text{if } \alpha = 1, \end{cases}
\end{aligned}$$

where

$$\mathrm{sgn}(u) = \begin{cases} 1 & \text{if } u > 0, \\ 0 & \text{if } u = 0, \\ -1 & \text{if } u < 0. \end{cases}$$

We say that η is strictly α-stable whenever $\mu = 0$. If, in addition $\beta = 0$, η is said to be symmetric α-stable. For an $\mathbb{R}$-valued normalized (standard) symmetric α-stable Lévy process $z(t)$, it has the characteristic function

$$\mathbb{E}\exp(iuz(t)) = e^{-t|u|^\alpha}, \quad u \in \mathbb{R}.$$

For more details on α-stable processes, we refer to Applebaum [3] and Sato [126]. Let $\mathscr{D} = D([-\tau, 0]; H)$ be the space of all càdlàg paths $f : [-\tau, 0] \mapsto H$ endowed with the Skorohod metric, and $(Z(t))_{t\geq 0}$ a cylindrical α-stable process defined by

$$Z(t) = \sum_{k=1}^{\infty} \beta_k Z_k(t) e_k,$$

where $(e_k)_{k\geq 1}$ is an orthogonal basis of H, $(Z_k(t))_{k\geq 1}$ is a sequence of i.i.d. $\mathbb{R}$-valued symmetric α-stable Lévy processes defined on $(\Omega, \mathscr{F}, (\mathscr{F}_t)_{t\geq 0}, \mathbb{P})$, and $(\beta_k)_{k\geq 1}$ is a sequence of positive numbers.

Consider a semi-linear FSPDE on the Hilbert space $(H, \langle\cdot,\cdot\rangle_H, \|\cdot\|_H)$

$$\mathrm{d}X(t) = \{AX(t) + b(X_t)\}\mathrm{d}t + \mathrm{d}Z(t), \quad t > 0, \quad X_0 = \xi \in \mathscr{D}, \tag{1.48}$$

where $b : \mathscr{D} \mapsto H$ is progressively measurable. Besides (A1), we assume the following conditions hold.

(C1) There exists an $L > 0$ such that

$$\|b(\phi) - b(\psi)\|_H \leq L \int_{[-\tau,0]} \|\phi(\theta) - \psi(\theta)\|_H \rho(\mathrm{d}\theta), \quad \phi, \psi \in \mathscr{D},$$

where $\rho(\cdot)$ is a probability measure on $[-\tau, 0]$.

(C2) $\delta := \sum_{k=1}^{\infty} \frac{\beta_k^\alpha}{\lambda_k} < \infty$ for some $\alpha \in (1, 2)$.

Under (A1) and (C1)–(C2), following the argument of [115, Theorem 5.3], (1.48) admits a unique mild solution, i.e., there exists a predictable H-valued stochastic process $(X(t;\xi))_{t\geq 0}$ such that

$$X(t;\xi) = e^{At}\xi(0) + \int_0^t e^{A(t-s)} b(X_s(\xi))\mathrm{d}s + \int_0^t e^{A(t-s)}\mathrm{d}Z(s).$$

Theorem 1.12 *Let* (A1) *and* (C1)–(C2) *hold and assume further* $L < \lambda_1$. *Then,* (1.48) *admits a unique invariant measure* $\pi(\cdot) \in \mathscr{P}(\mathscr{D})$.

Proof Let

$$\widetilde{Z}(t) = \sum_{k=1}^{\infty} \beta_k \widetilde{Z}_k(t) e_k,$$

where $(\widetilde{Z}_k(t))$ is an independent copy of $(Z_k(t))_{t\geq 0}$, and $(\overline{Z}(t))_{t\geq 0}$ be a double-sided cylindrical α-stable process defined by

$$\overline{Z}(t) := \begin{cases} Z(t), & t \geq 0, \\ \widetilde{Z}(-t), & t < 0 \end{cases}$$

with the filtration

$$\overline{\mathscr{F}}_t := \bigcap_{s>t} \overline{\mathscr{F}}^0_s,$$

where $\overline{\mathscr{F}}^0_s := \sigma(\{\overline{Z}(r_2) - \overline{Z}(r_1) : -\infty < r_1 \le r_2 \le s, \Gamma\}, \mathscr{N})$ and $\mathscr{N} := \{A \in \mathscr{F} | \mathbb{P}(A) = 0\}$. Fix $s \in \mathbb{R}$ and consider the following semi-linear FSPDE

$$\mathrm{d}X(t) = \{AX(t) + b(X_t)\}\mathrm{d}t + \mathrm{d}\overline{Z}(t), \quad t \ge s, \quad X_s = \xi. \tag{1.49}$$

Then (1.49) has a unique mild solution $X(t; s, \xi)$ with the initial data $X_s = \xi$. Set

$$\Upsilon(t, s) := \int_s^t e^{(t-u)A}\mathrm{d}\overline{Z}(u), \ t \ge s, \ \Upsilon(t, s) :\equiv 0, \ t < s, \ \Lambda(t, s) := X(t; s, \xi) - \Upsilon(t, s).$$

Then, (1.49) can be reformulated as

$$\mathrm{d}\Lambda(t, s) = \{A\Lambda(t, s) + b(\Lambda_t(s) + \Upsilon_t(s))\}\mathrm{d}t, \quad t > s, \quad \Lambda_s = \xi.$$

Using $\Gamma(t)$ defined in (1.5), and following the argument in the proof of Theorem 1.11, to complete the proof of Theorem 1.12, we need only show that

$$\mathbb{E}\|\Gamma_t\|_\infty \lesssim \left(1 + \|\xi(0)\|_H + \int_{-\tau}^0 \|\xi(\theta)\|_H \mathrm{d}\theta\right)e^{-\lambda(t-s_2)}, \tag{1.50}$$

and that

$$\mathbb{E}\|X_t(\xi) - X_t(\eta)\|_\infty \lesssim \left(|\xi(0) - \eta(0)| + \int_{-\tau}^0 |\xi(\theta) - \eta(\theta)|\mathrm{d}\theta\right)e^{-\lambda t} \tag{1.51}$$

for some $\lambda > 0$. To proceed, we verify that (1.50) and (1.51) hold, respectively. For any $\nu > 0$ and arbitrary $\varepsilon > 0$, by the chain rule, we deduce from (A1) that

$$\begin{aligned} &e^{\nu t}\sqrt{(\varepsilon + \|\Lambda(t, s)\|_H^2)} \\ &\le e^{\nu s}\sqrt{(\varepsilon + \|\xi(0)\|_H^2)} + \lambda_1\varepsilon^{1/2}\int_s^t e^{\nu u}\mathrm{d}u \\ &+ \int_s^t e^{\nu u}\left\{(\nu - \lambda_1)\sqrt{(\varepsilon + \|\Lambda(u, s)\|_H^2)} + \|b(\Lambda_u(s) + \Upsilon_u(s))\|_H\right\}\mathrm{d}u. \end{aligned} \tag{1.52}$$

Letting $\varepsilon \to 0$ in (1.52) and using (C1) give that

$$\begin{aligned}
&e^{\nu t}\|\Lambda(t,s)\|_H \\
&\le e^{\nu s}\|\xi(0)\|_H + \int_s^t e^{\nu u}\{(\nu-\lambda_1)\|\Lambda(u,s)\|_H + \|b(\Lambda_u(s)+\Upsilon_u(s))\|_H\}\mathrm{d}u \\
&\le e^{\nu s}\|\xi(0)\|_H + ce^{\nu t} - (\lambda_1-\nu-Le^{\nu\tau})\int_s^t e^{\nu u}\|\Lambda(u,s)\|_H\mathrm{d}u \qquad (1.53)\\
&\quad + Le^{\nu\tau}\int_s^t e^{\nu u}\|\Upsilon(u,s)\|_H\mathrm{d}u + Le^{\nu\tau}\int_{s-\tau}^s e^{\nu u}\|\xi(u-s)\|_H\mathrm{d}u.
\end{aligned}$$

Next, since

$$\mathscr{L}\Big(\int_s^t e^{-\lambda_k(t-r)}\beta_k\mathrm{d}\widetilde{Z}_k(r)\Big) = \mathscr{L}\Big(\int_0^{t-s} e^{-\lambda_k(t-s-r)}\beta_k\mathrm{d}\widehat{Z}_k(r)\Big),$$

where $\widehat{Z}_k(r) := \widetilde{Z}_k(r+s) - \widetilde{Z}_k(s)$ is the shift version of $\widetilde{Z}_k(r)$, we obtain from [115, (4.12)] and (C2) that

$$\mathbb{E}\|\Upsilon(t,s)\|_H \lesssim \Big(\sum_{k=1}^\infty \beta_k^\alpha\frac{1-e^{-\alpha\lambda_k(t-s)}}{\alpha\lambda_k}\Big)^{1/\alpha} < \infty. \qquad (1.54)$$

This, together with (1.53), leads to

$$\sup_{t\ge s}\mathbb{E}\|\Lambda(t,s)\|_H \lesssim 1 + \|\xi(0)\|_H + \int_{-\tau}^0 \|\xi(\theta)\|_H\mathrm{d}\theta \qquad (1.55)$$

by choosing $\nu \in (0,\lambda_1)$ sufficiently small such that $\lambda_1 - \nu - Le^{\nu\tau} > 0$ due to $\lambda_1 > L$. For $-\infty < s_1 \le s_2 \le t < \infty$, following the procedure to derive (1.52), we deduce that

$$\begin{aligned}
&\mathbb{E}\Big(\sup_{t-\tau\le r\le t}(e^{\nu r}\|\Lambda(r,s_1)-\Lambda(r,s_2)\|_H)\Big) \\
&\le e^{\nu s_2}\mathbb{E}\|\Lambda(s_2,s_1)-\xi(0)\|_H + e^{\nu\tau}L\int_{s_2}^t e^{-(\lambda_1-\nu)u+\lambda_1 s_2}\mathrm{d}u\,\mathbb{E}\Big\|\int_{s_1}^{s_2} e^{(s_2-r)A}\mathrm{d}\overline{Z}(r)\Big\|_H \\
&\quad + e^{\nu\tau}L\int_{s_2-\tau}^{s_2} e^{\nu u}\mathbb{E}\|\Gamma(u)\|_H\mathrm{d}u + 2e^{\nu\tau}L\int_{s_2-\tau}^{s_2} e^{\nu u}\mathbb{E}\|\Upsilon(u,s_1)\|_H\mathrm{d}u.
\end{aligned}$$

This, together with (1.54) and (1.55), yields that

$$\mathbb{E}\|\Lambda_t(s_1)-\Lambda_t(s_2)\|_\infty \lesssim \Big(1+\|\xi(0)\|_H + \int_{-\tau}^0\|\xi(\theta)\|_H\mathrm{d}\theta\Big)e^{-\nu(t-s_2)}. \qquad (1.56)$$

Moreover, from (1.54) and (1.56), we have

$$\begin{aligned}\mathbb{E}\|\Gamma_t\|_\infty &\le \mathbb{E}\|\Lambda_t(s_1) - \Lambda_t(s_2)\|_\infty + \Big(\sup_{t-\tau\le r\le t} \|e^{(r-s_2)A}\|\Big)\mathbb{E}\|\Upsilon(s_2, s_1)\|_H \\ &\lesssim \Big(1 + \|\xi(0)\|_H + \int_{-\tau}^0 \|\xi(\theta)\|_H \mathrm{d}\theta\Big)e^{-\nu(t-s_2)}.\end{aligned}$$

Thus, (1.50) follows. Next, (1.51) can be proved by carrying out the argument similar to that of deriving (1.56).

Chapter 2
Ergodicity for Functional Stochastic Equations Without Dissipativity

Dealing with diffusions with "pure delay" in which both the drift and the diffusion coefficients depend only on the arguments with delays, most of the existing results are not applicable. This chapter uses variation-of-constants formulae to overcome the difficulties due to the lack of the information at the current time, and establishes existence and uniqueness of invariant measures for functional stochastic equations that need not satisfy dissipative conditions. The functional stochastic equations considered in this chapter cover FSDEs of neutral type, FSPDEs driven by cylindrical Wiener processes, and FSDEs driven by Lévy processes without finite second moments. Sections 2.1–2.3 and 2.5 of this chapter are mainly based on the results in [13].

2.1 Introduction

One of the main approaches in the literature to date is to incorporate certain dissipativity to establish existence of invariant measures for functional stochastic equations. The dissipativity is normally assured by imposing conditions of the current time with certain decay conditions. Such an idea has been used extensively. For instance, utilizing the remote method (i.e., the dissipative method), Bao et al. [12] treated several classes of functional stochastic equations; applying an exponential-type estimate, Bo-Yuan [21] investigated stochastic differential delay equations (SDDEs) driven by Poisson jump processes; adopting the Arzelà–Ascoli tightness characterization, Es-Sarhir et al. [54] and Kinnally-Williams [79] considered FSDEs with super-linear drift terms and positivity constraints, respectively. Nevertheless, the existing literature is unable to deal with the following seemingly simple linear SDDE on the real line $\mathbb{R}$,

$$\mathrm{d}X(t) = -X(t-1)\mathrm{d}t + \sigma X(t-1)\mathrm{d}W(t), \quad X_0 = \xi, \tag{2.1}$$

J. Bao et al., *Asymptotic Analysis for Functional Stochastic Differential Equations*, SpringerBriefs in Mathematics, DOI 10.1007/978-3-319-46979-9_2

where $\sigma \in \mathbb{R}$ and $(W(t))_{t\geq 0}$ is a real-valued standard Brownian motion. Observe that it is impossible to choose constants $\lambda_1 > \lambda_2 > 0$ such that

$$-2xy + \sigma^2 y^2 \leq -\lambda_1 x^2 + \lambda_2 y^2, \quad \forall\ x, y \in \mathbb{R} \tag{2.2}$$

holds even for some $\sigma \in \mathbb{R}$ sufficiently small. Indeed, if (2.2) holds, then

$$\lambda_1 \leq \beta\Big(\frac{y}{x} + \frac{1}{\beta}\Big)^2 - \frac{1}{\beta}, \quad x \neq 0, \tag{2.3}$$

where $\beta := \lambda_2 - \sigma^2 > 0$. It is readily seen that (2.3) no longer holds provided that $-\frac{2}{\beta} \leq \frac{y}{x} \leq 0$. That is, (2.1) does not obey a dissipative condition. Therefore, the techniques used in [12, 21, 54, 79] are not applicable to (2.1). We remark that the main problem is due to the lack of the information at the current time, so no dissipative conditions can be used.

In the well-known work [153], Yorke treated deterministic systems with pure delays. His work has stimulated much of the subsequent work resulting in a vast literature on the pure delay equations in the deterministic setup; see also Wu et al. [144] for a substantial extension to systems with Markov switching. Our consideration in this chapter is a generalization of the model in [153] in which both the drift and diffusion coefficients may involve only retarded elements. The right-hand sides of such differential equations may not involve information on the current time. Consequently, it is impossible to use any dissipative conditions.

In this chapter, we aim to obtain existence and uniqueness of invariant measures for several classes of functional stochastic equations, which include (2.1) as a special case. The crucial point is that we focus on functional stochastic systems without satisfying dissipative conditions. To overcome the difficulties, we use the variation-of-constants formulae, which have been applied successfully in [5, 60, 95, 119]. More precisely, [5] examined the exact almost sure growth rate of the running maximum of solutions for affine FSDEs, whereas in [60] and [95], the authors considered stationary solutions (see Remark 2.7) for finite-dimensional and infinite-dimensional retarded Ornstein–Uhlenbeck (O–U) processes, respectively. It is also worth pointing out that the variation-of-constants formula together with the semi-martingale characteristics has been utilized to study existence of invariant measures for a class of FSDEs driven by jump processes [119], which do not include (2.1), however.

The remainder of this chapter is organized as follows. Section 2.2 is devoted to existence and uniqueness of invariant measures for semi-linear FSDEs driven by Brownian motions using the Arzelà–Ascoli type tightness characterization. Sections 2.3 and 2.4 extend the main result derived in Section 2.2 to FSDEs of neutral type and FSPDEs driven by cylindrical Wiener processes, respectively. Section 2.5 focuses on existence and uniqueness of invariant measures for semi-linear FSDEs driven by Lévy processes, which need not have finite second moments. The key is to use the variation-of-constants formula together with the tightness criterion due to Kurtz [55].

2.2 Ergodicity for FSDEs Driven by Brownian Motions

Let μ be an $\mathbb{R}^n \otimes \mathbb{R}^n$-valued finite signed measure on $[-\tau, 0]$; that is, $\mu = (\mu_{ij})_{1\le i,j\le n}$, where each μ_{ij} is a finite signed measure on $[-\tau, 0]$. Consider the following semi-linear FSDE

$$\mathrm{d}X(t) = \Big(\int_{[-\tau,0]} \mu(\mathrm{d}\theta)X(t+\theta)\Big)\mathrm{d}t + \sigma(X_t)\mathrm{d}W(t), \quad t > 0, \quad X_0 = \xi. \tag{2.4}$$

Herein, $\sigma : \mathscr{C} \mapsto \mathbb{R}^n \otimes \mathbb{R}^m$ satisfies (H2) in Chapter 1. By virtue of [99, Theorem 2.2, p. 150], (2.4) admits a unique strong solution $(X(t;\xi))_{t\ge -\tau}$ with the initial data $X_0 = \xi \in \mathscr{C}$. In what follows, we often write $X(t)$ and X_t in lieu of $X(t;\xi)$ and $X_t(\xi)$, respectively, for notational simplicity.

Let $\mathbb{C}$ be the set of all complex numbers and $\mathrm{Re}(z)$ the real part of $z \in \mathbb{C}$. Set

$$v_\mu := \sup\{\mathrm{Re}(\lambda) : \lambda \in \mathbb{C}, \Delta_\mu(\lambda) = 0\}. \tag{2.5}$$

Herein,

$$\Delta_\mu(\lambda) := \det\Big(\lambda I_{n\times n} - \int_{[-\tau,0]} e^{\lambda s}\mu(\mathrm{d}s)\Big),$$

where $\det(A)$ denotes the determinant of $A \in \mathbb{R}^n \otimes \mathbb{R}^n$, and $I_{n\times n} \in \mathbb{R}^n \otimes \mathbb{R}^n$ is the unitary matrix. $\Delta_\mu(\lambda) = 0$ is called the characteristic equation associated with the deterministic counterpart of (2.4) (i.e., $\sigma \equiv \mathbf{0}_{n\times m}$ therein, where $\mathbf{0}_{n\times m}$ is the $n \times m$ zero matrix).

In particular, when $\mu = A\delta_0$ with $A \in \mathbb{R}^n \otimes \mathbb{R}^n$ and δ_0 being the Dirac delta measure at the point 0, equation (2.4) reduces to the semi-linear FSDE:

$$\mathrm{d}X(t) = AX(t)\mathrm{d}t + \sigma(X_t)\mathrm{d}W(t), \quad t > 0, \quad X_0 = \xi,$$

and v_μ is the largest real part of eigenvalues of A. Suppose that $(\Gamma(t))_{t\ge 0}$ solves the following $\mathbb{R}^n \otimes \mathbb{R}^n$-valued equation

$$\mathrm{d}\Gamma(t) = \Big(\int_{[-\tau,0]} \mu(\mathrm{d}\theta)\Gamma(t+\theta)\Big)\mathrm{d}t \tag{2.6}$$

with the initial data $\Gamma(0) = I_{n\times n}$, $\Gamma(\theta) = \mathbf{0}_{n\times n}$ for $\theta \in [-\tau, 0)$. For more properties of Γ, please refer to Theorem B.3. The following lemma provides a variation-of-constants formula for (2.4).

Lemma 2.1 *Equation* (2.4) *admits a unique strong solution* $(X(t;\xi))_{t\ge -\tau}$, *which can be represented by*

$$X(t;\xi) = \Gamma(t)\xi(0) + \int_{[-\tau,0]} \mu(d\theta) \int_\theta^0 \Gamma(t+\theta-s)\xi(s)ds + \int_0^t \Gamma(t-s)\sigma(X_s(\xi))dW(s), \quad t > 0. \tag{2.7}$$

Proof Let $t \geq 0$ be fixed. By the chain rule, it follows that

$$\mathrm{d}(\Gamma(t-s)X(s)) = \mathrm{d}(\Gamma(t-s))X(s) + \Gamma(t-s)\mathrm{d}X(s), \quad s \in [0,t].$$

Integrating from 0 to t, we deduce from (2.4) and (2.6) that

$$\begin{aligned}
&X(t) - \Gamma(t)\xi(0)\\
&= -\int_0^t \Big(\int_{[-\tau,0]} \mu(\mathrm{d}\theta)\Gamma(t-s+\theta)\Big)X(s)\mathrm{d}s\\
&\quad + \int_0^t \Gamma(t-s)\Big(\int_{[-\tau,0]} \mu(\mathrm{d}\theta)X(s+\theta)\Big)\mathrm{d}s + \int_0^t \Gamma(t-s)\sigma(X_s)\mathrm{d}W(s)\\
&= -\int_{[-\tau,0]} \mu(\mathrm{d}\theta) \int_0^t \Gamma(t-s+\theta)X(s)\mathrm{d}s\\
&\quad + \int_{[-\tau,0]} \mu(\mathrm{d}\theta) \int_\theta^{t+\theta} \Gamma(t-s+\theta)X(s)\mathrm{d}s + \int_0^t \Gamma(t-s)\sigma(X_s)\mathrm{d}W(s)\\
&= \int_{[-\tau,0]} \mu(\mathrm{d}\theta) \int_\theta^0 \Gamma(t-s+\theta)X(s)\mathrm{d}s - \int_{[-\tau,0]} \mu(\mathrm{d}\theta) \int_{t+\theta}^t \Gamma(t-s+\theta)X(s)\mathrm{d}s\\
&\quad + \int_0^t \Gamma(t-s)\sigma(X_s)\mathrm{d}W(s)\\
&= \int_{[-\tau,0]} \mu(\mathrm{d}\theta) \int_\theta^0 \Gamma(t+\theta-s)\xi(s)\mathrm{d}s + \int_0^t \Gamma(t-s)\sigma(X_s)\mathrm{d}W(s),
\end{aligned}$$

where in the second step we have used substitution, and in the last two steps utilized additive property of integral together with the observation $\Gamma(0) = I_{n\times n}, \Gamma(\theta) = \mathbf{0}_{n\times n}, \theta \in [-\tau, 0)$. Thus, (2.7) follows.

Remark 2.2 With regard to the variation-of-constants formula for semi-linear FSDEs driven by a general semi-martingale, we refer to Reiß et al. [118]. While, in Lemma 2.1, we provide an alternative proof for the variation-of-constants formula (2.7) to make the content self-contained.

In what follows, we assume $v_\mu < 0$. By Theorem B.3, for any $\lambda \in (0, -v_\mu)$, there exists a constant $c_\lambda > 0$ such that

$$\|\Gamma(t)\| \leq c_\lambda e^{-\lambda t}, \quad t \geq -\tau, \tag{2.8}$$

where $\|\Gamma(t)\|$ denotes the operator norm of the matrix $\Gamma(t) \in \mathbb{R}^n \otimes \mathbb{R}^n$. We remark that the optimal constant c_λ is increasing for $\lambda \in (0, -v_\mu)$. If, in particular, $\mu = A\delta_0$ for a symmetric $n \times n$-matrix A, (2.8) holds with $c_\lambda = 1$ and $\lambda \in (0, -v_\mu]$. In general, see Theorem B.5 in the Appendix B.2 for an upper bound of c_λ.

The following lemma plays a crucial role in obtaining the long-term behavior of segment process $(X_t(\xi))_{t\geq 0}$.

Lemma 2.3 *Let $(u(t))_{t\geq 0}$ be an $\mathbb{R}^n \otimes \mathbb{R}^m$-valued predictable process. Then, for any $p \geq 1$ and $t > 0$,*

$$\left(\mathbb{E}\left|\int_0^t \Gamma(t-s)u(s)dW(s)\right|^{2p}\right)^{1/p} \leq p(2p-1)\int_0^t (\mathbb{E}\|\Gamma(t-s)u(s)\|_{HS}^{2p})^{1/p}ds$$

provided that the integral on the right-hand side is finite for each $t > 0$.

Proof Observe that this estimate cannot be obtained directly from [36, Lemma 7.7, p. 174] due to the fact that $\int_0^t \Gamma(t-s)u(s)\mathrm{d}W(s)$ is not a martingale. Nevertheless, some ideas of the aforementioned reference can be used. To make it self-contained, we provide an outline of the proof. Let

$$M(r) = \int_0^r \Gamma(t-s)u(s)\mathrm{d}W(s), \quad r \in [0, t).$$

In what follows, we let $r \in [0, t)$. By Itô's formula and then Hölder's inequality, it follows that

$$\begin{aligned}\mathbb{E}|M(r)|^{2p} &\leq p(2p-1)\int_0^r \mathbb{E}(|M(s)|^{2(p-1)}\|\Gamma(t-s)u(s)\|_{HS}^2)\mathrm{d}s\\ &\leq p(2p-1)\sup_{0\leq s\leq r}(\mathbb{E}|M(s)|^{2p})^{(p-1)/p}\int_0^r (\mathbb{E}\|\Gamma(t-s)u(s)\|_{HS}^{2p})^{1/p}\mathrm{d}s.\end{aligned}$$

Hence, one has

$$\sup_{0\leq s\leq r}\mathbb{E}|M(s)|^{2p} \leq p(2p-1)\sup_{0\leq s\leq r}(\mathbb{E}|M(s)|^{2p})^{(p-1)/p}\int_0^r (\mathbb{E}\|\Gamma(t-s)u(s)\|_{HS}^{2p})^{1/p}\mathrm{d}s$$

owing to the fact that $\sup_{0\leq s\leq r}(\mathbb{E}|M(s)|^{2p})^{(p-1)/p}$ is nondecreasing w.r.t. r. This further implies that

$$\begin{aligned}\sup_{0\leq s\leq r}(\mathbb{E}|M(s)|^{2p})^{1/p} &\leq p(2p-1)\int_0^r (\mathbb{E}\|\Gamma(t-s)u(s)\|_{HS}^{2p})^{1/p}\mathrm{d}s\\ &\leq p(2p-1)\int_0^t (\mathbb{E}\|\Gamma(t-s)u(s)\|_{HS}^{2p})^{1/p}\mathrm{d}s.\end{aligned}$$

Consequently, the desired assertion follows by taking $r \uparrow t$.

The lemma below gives some sufficient conditions to guarantee that (2.4) admits a unique invariant measure.

Lemma 2.4 *Let* $v_\mu < 0$ *and assume further that* (H2) *holds with* $\lambda_3 > 0$ *such that* $c_\lambda^2\lambda_3(1+e^{2\lambda\tau}\rho([-\tau,0])) < \lambda$ *for* $\lambda \in (0, -v_\mu)$*, where* $c_\lambda > 0$ *is introduced in* (2.8). *Then there exist* $p > 1$ *and* $\lambda' > 0$ *such that*

$$\sup_{t\geq 0}\mathbb{E}\|X_t(\xi)\|_\infty^{2p} \lesssim 1 + \|\xi\|_\infty^{2p}, \tag{2.9}$$

and

$$\mathbb{E}\|X_t(\xi) - X_t(\eta)\|_\infty^2 \lesssim e^{-\lambda' t}\|\xi - \eta\|_\infty^2. \tag{2.10}$$

Proof Owing to

$$c_\lambda^2\lambda_3(1 + e^{2\lambda\tau}\rho([-\tau,0])) < \lambda \text{ for } \lambda \in (0, -v_\mu),$$

we can choose $\kappa, p > 1$ and $\epsilon > 0$ such that

$$\beta_{\kappa,\epsilon}(p) := p(2p-1)(1+\epsilon)\kappa^{1/p}c_\lambda^2\lambda_3(1 + e^{2\lambda\tau}\rho([-\tau,0])) < \lambda. \tag{2.11}$$

In what follows, we fix $\kappa, p > 1$ and $\varepsilon > 0$ such that (2.11) holds. Then there exists a constant $\gamma > 1$ such that $(a+b+c)^{2p} \leq \kappa a^{2p} + \gamma(b^{2p} + c^{2p})$ with $a, b, c > 0$. Thus, by the elementary inequality $(a+b)^\theta \leq a^\theta + b^\theta$ for $a, b \geq 0$ and $\theta \in (0, 1)$, it follows from (2.7) and (2.8) that

$$\begin{aligned}(\mathbb{E}|X(t)|^{2p})^{1/p} &\leq \kappa^{1/p}\Theta_p(t) + \gamma^{1/p}\Big\{|\Gamma(t)\xi(0)|^2 \\ &\quad + \Big|\int_{[-\tau,0]}\mu(\mathrm{d}\theta)\int_\theta^0 \Gamma(t+\theta-s)\xi(s)\mathrm{d}s\Big|^2\Big\} \\ &\leq ce^{-2\lambda t}\|\xi\|_\infty^2 + \kappa^{1/p}\Theta_p(t),\end{aligned} \tag{2.12}$$

where

$$\Theta_p(t) := \Big(\mathbb{E}\Big|\int_0^t \Gamma(t-s)\sigma(X_s)\mathrm{d}W(s)\Big|^{2p}\Big)^{1/p}.$$

Set $\alpha(p) := p(2p-1)c_\lambda^2\lambda_3$. Applying Lemma 2.3 with $u(s) = \sigma(X_s)$, followed by using (2.8), we derive from (H2) that

$$\begin{aligned}\Theta_p(t) &\leq \alpha(p)\int_0^t e^{-2\lambda(t-s)}\Big\{c + \Big(\mathbb{E}\Big(|X(s)|^2 + \int_{[-\tau,0]}|X(s+\theta)|^2\nu(\mathrm{d}\theta)\Big)^p\Big)^{1/2p}\Big\}^2\mathrm{d}s \\ &\leq \alpha(p)\int_0^t e^{-2\lambda(t-s)}\Big\{c + (1+\varepsilon)\Big(\mathbb{E}\Big(|X(s)|^2 + \int_{[-\tau,0]}|X(s+\theta)|^2\nu(\mathrm{d}\theta)\Big)^p\Big)^{1/p}\Big\}\mathrm{d}s \\ &\leq \alpha(p)\int_0^t e^{-2\lambda(t-s)}\Big\{c + (1+\varepsilon)\Big((\mathbb{E}|X(s)|^{2p})^{1/p}\end{aligned}$$

$$+\int_{[-\tau,0]}(\mathbb{E}|X(s+\theta)|^{2p})^{1/p}\nu(\mathrm{d}\theta)\Big)\Big\}\mathrm{d}s$$
$$\le c\Big(\int_0^t e^{-2\lambda(t-s)}\mathrm{d}s+e^{-2\lambda t}\|\xi\|_\infty^2\Big)$$
$$+2\alpha(p)(1+\varepsilon)(1+e^{2\lambda\tau}\rho([-\tau,0]))\int_0^t e^{-2\lambda(t-s)}\left(\mathbb{E}|X(s)|^{2p}\right)^{1/p}\mathrm{d}s$$

where in the first step we have used Minkovskii's inequality [99, p. 5], and in the second step utilized the elementary inequality: $(a+b)^2\le(1+\theta)a^2+(1+\frac{1}{\theta})b^2$ with $\theta=\varepsilon$. Inserting the previous estimate on $\Theta_p(t)$ into (2.12) yields that

$$e^{2\lambda t}(\mathbb{E}|X(t)|^{2p})^{1/p}\le c_\varepsilon\Big(\int_0^t e^{2\lambda s}\mathrm{d}s+\|\xi\|_\infty^2\Big)+2\beta_{\kappa,\varepsilon}(p)\int_0^t e^{2\lambda s}(\mathbb{E}|X(s)|^{2p})^{1/p}\mathrm{d}s,$$

where $\beta_{\kappa,\varepsilon}(p)>0$ is defined in (2.11). Thus, by Gronwall's inequality, we have

$$e^{2\lambda t}(\mathbb{E}|X(t)|^{2p})^{1/p}\le c_\varepsilon\Big(\int_0^t e^{2\lambda s}\mathrm{d}s+\|\xi\|_\infty^2\Big)$$
$$+2c_\varepsilon\beta_{\kappa,\varepsilon}(p)\int_0^t\Big(\int_0^s e^{2\lambda u}\mathrm{d}u+\|\xi\|_\infty^2\Big)e^{2\beta_{\kappa,\varepsilon}(p)(t-s)}\mathrm{d}s.$$

Hence, by interchanging the order of integral, we obtain that

$$\begin{aligned}(\mathbb{E}|X(t)|^{2p})^{1/p}&\lesssim 1+\|\xi\|_\infty^2+\int_0^t\int_0^s e^{-2\lambda(t-u)}e^{2\beta_{\kappa,\varepsilon}(p)(t-s)}\mathrm{d}u\mathrm{d}s\\&\quad+\|\xi\|_\infty^2e^{-2\lambda t}\int_0^t e^{2\beta_{\kappa,\varepsilon}(p)(t-s)}\mathrm{d}s\\&\lesssim 1+\|\xi\|_\infty^2+\int_0^t e^{-2\lambda(t-u)}\int_u^t e^{2\beta_{\kappa,\varepsilon}(p)(t-s)}\mathrm{d}s\mathrm{d}u\\&\quad+\|\xi\|_\infty^2e^{-2\lambda t}\int_0^t e^{2\beta_{\kappa,\varepsilon}(p)(t-s)}\mathrm{d}s\\&\lesssim 1+\|\xi\|_\infty^2,\end{aligned}\tag{2.13}$$

where in the last step we have used $\beta_{\kappa,\varepsilon}(p)<\lambda$ due to (2.11). Next, applying Hölder's inequality for integrals with respect to time and B–D–G's inequality for the term involving martingale, we deduce from (H2) and (2.13) that

$$
\begin{aligned}
\mathbb{E}\|X_t\|_\infty^{2p} &\lesssim \mathbb{E}|X(t-\tau)|^{2p} + \int_{t-2\tau}^t \mathbb{E}|X(s)|^{2p}\mathrm{d}s + \mathbb{E}\Big(\int_{t-\tau}^t \|\sigma(X_u)\|_{HS}^2\mathrm{d}u\Big)^p \\
&\lesssim 1 + \mathbb{E}|X(t-\tau)|^{2p} + \int_{t-2\tau}^t \mathbb{E}|X(s)|^{2p}\mathrm{d}s \\
&\lesssim 1 + \|\xi\|_\infty^{2p}, \quad t \geq \tau.
\end{aligned} \tag{2.14}
$$

Similarly,

$$
\mathbb{E}\Big(\sup_{0\leq t\leq \tau} |X(t)|^{2p}\Big) \lesssim 1 + \|\xi\|_\infty^{2p}. \tag{2.15}
$$

As a result, (2.9) follows from (2.14) and (2.15) immediately.

Next, we show (2.10). Let $\Theta(t) = X(t,\xi) - X(t,\eta)$ for notational simplicity. Using the elementary inequality (1.21), followed by utilizing Hölder's inequality for the time integrals and Itô's isometry for the term involving martingale, we obtain that

$$
\begin{aligned}
&\mathbb{E}|\Theta(t)|^2 \\
&\leq \frac{1+\alpha}{\alpha}\Big|\Gamma(t)(\xi(0)-\eta(0)) + \int_{[-\tau,0]}\mu(\mathrm{d}\theta)\int_\theta^0 \Gamma(t+\theta-s)(\xi(s)-\eta(s))\mathrm{d}s\Big|^2 \\
&\quad + (1+\alpha)\int_0^t \mathbb{E}\|\Gamma(t-s)(\sigma(X_s(\xi))-\sigma(X_s(\eta)))\|_{HS}^2\mathrm{d}s \\
&\lesssim \frac{1+\alpha}{\alpha}e^{-2\lambda t}\|\xi-\eta\|_\infty^2 \\
&\quad + (1+\alpha)c_\lambda^2\lambda_3\int_0^t e^{-2\lambda(t-s)}\Big(|\Theta(s)| + \int_{[-\tau,0]}|\Theta(s+\theta)|\rho(\mathrm{d}\theta)\Big)^2\mathrm{d}s \\
&\lesssim \frac{1+\alpha}{\alpha}e^{-2\lambda t}\|\xi-\eta\|_\infty^2 + 2(1+\alpha)c_\lambda^2\lambda_3\int_0^t e^{-2\lambda(t-s)}\Big(|\Theta(s)|^2 \\
&\quad + \rho([-\tau,0])\int_{[-\tau,0]}|\Theta(s+\theta)|^2\rho(\mathrm{d}\theta)\Big)\mathrm{d}s \\
&\lesssim c_\alpha e^{-2\lambda t}\|\xi-\eta\|_\infty^2 + 2\beta(\alpha)\int_0^t e^{-2\lambda(t-s)}\mathbb{E}|\Theta(s)|^2\mathrm{d}s,
\end{aligned}
$$

where $\beta(\alpha) := (1+\alpha)c_\lambda^2\lambda_3(1+e^{2\lambda\tau}\rho([-\tau,0]))$. So we arrive at

$$
e^{2\lambda t}\mathbb{E}|\Theta(t)|^2 \lesssim c_\alpha\|\xi-\eta\|_\infty^2 + 2\beta(\alpha)\int_0^t e^{2\lambda s}\mathbb{E}|\Theta(s)|^2\mathrm{d}s.
$$

Since $c_\lambda^2\lambda_3(1+e^{2\lambda\tau}\rho([-\tau,0])) < \lambda$, we can choose $\alpha > 0$ such that $\beta(\alpha) < \lambda$ holds. Thus, Gronwall's inequality leads to

$$
\mathbb{E}|\Theta(t)|^2 \lesssim e^{-2(\lambda-\beta(\alpha))t}\|\xi-\eta\|_\infty^2.
$$

Finally, (2.10) follows by carrying out the argument for deriving (2.14).

Theorem 2.5 *Assume that the assumptions of Lemma 2.4 hold. Then,* (2.4) *has a unique invariant measure* π, *which is exponentially mixing.*

Proof The proof on existence of an invariant measure is facilitated by use of the classical Arzelà–Ascoli theorem [20, Theorem 8.3, p. 56]. For any integer $q \geq 1$, set

$$\mu_q(\cdot) := \frac{1}{q}\int_0^q \mathbb{P}(t, \xi, \cdot)\mathrm{d}t,$$

where $\mathbb{P}(t, \xi, \cdot)$ is the Markovian transition kernel of $X_t(\xi)$. By virtue of the Krylov–Bogoliubov theorem (see, e.g., [37, Theorem 3.1.1, p. 21]), to establish the existence of an invariant measure, it is sufficient to verify that $(\mu_q(\cdot))_{q\geq 1}$ is relatively compact. Note that the phase space $\mathscr{C}$ for the segment process $(X_t(\xi))_{t\geq 0}$ is a complete separable metric space under the uniform metric $\|\cdot\|_\infty$ (see, e.g., [20, p. 220]). Taking [20, Theorem 6.2, p. 37] into consideration, we need only prove that $(\mu_q(\cdot))_{q\geq 1}$ is tight. Moreover, by virtue of [20, Theorem 8.2, p. 55], it suffices to verify that

$$\lim_{\delta\downarrow 0}\sup_{q\geq 1}\mu_q(\varphi\in\mathscr{C} : w_{[-\tau,0]}(\varphi, \delta) \geq \varepsilon) = 0 \tag{2.16}$$

for any $\varepsilon > 0$, where $w_{[-\tau,0]}(\varphi, \delta)$, the modulus of continuity of φ, is defined by

$$w_{[-\tau,0]}(\varphi, \delta) = \sup_{|s-t|\leq\delta, s,t\in[-\tau,0]} |\varphi(s) - \varphi(t)|, \quad \delta > 0.$$

Noting that

$$\begin{aligned} I(t, \delta) :&= \sup_{t\leq v\leq u\leq t+\tau, 0\leq u-v\leq\delta} |X(u) - X(v)| \\ &\leq \sup_{t\leq v\leq u\leq t+\tau, 0\leq u-v\leq\delta} \int_v^u \left|\int_{[-\tau,0]}\mu(\mathrm{d}\theta)X(s+\theta)\right|\mathrm{d}s \\ &\quad + \sup_{t\leq v\leq u\leq t+\tau, 0\leq u-v\leq\delta}\left|\int_v^u \sigma(X_s)\mathrm{d}W(s)\right| \\ &=: I_1(t, \delta) + I_2(t, \delta), \quad t \geq \tau, \end{aligned}$$

one has

$$\mathbb{P}(I(t, \delta) \geq \varepsilon) \leq \mathbb{P}(I_1(t, \delta) \geq \varepsilon/2) + \mathbb{P}(I_2(t, \delta) \geq \varepsilon/2).$$

Next, by virtue of Chebyshev's inequality and (2.9),

$$\mathbb{P}(I_1(t, \delta) \geq \varepsilon/2) \lesssim \varepsilon^{-2}\mathbb{E}I_1(t, \delta)^2 \lesssim (\mathbb{E}\|X_t\|_\infty^2 + \mathbb{E}\|X_{t+\tau}\|_\infty^2)\delta^2 \lesssim \delta^2. \tag{2.17}$$

On the other hand, for $p > 1$ satisfying (2.11) and arbitrary $0 \le s \le t$, by Lemma [99, Theorem 7.1, p. 39], together with (H2) and (2.9), it follows that

$$\mathbb{E}\Big|\int_s^t \sigma(X_r)\mathrm{d}W(r)\Big|^{2p} \lesssim (t-s)^{p-1}\int_s^t \Big\{1+\sup_{u\ge 0}\mathbb{E}\|X_u\|_\infty^{2p}\Big\}\mathrm{d}r \lesssim (t-s)^p.$$

This, combined with the Kolmogorov tightness criterion (see, e.g., [77, Problem 4.11, p. 64]), implies that

$$\lim_{\delta\downarrow 0}\sup_{t\ge\tau}\mathbb{P}(I_2(t,\delta)\ge\varepsilon/2)=0. \tag{2.18}$$

Consequently, (2.16) follows from (2.17), (2.18), and

$$\mu_q(\varphi\in\mathscr{C}: w_{[-\tau,0]}(\varphi,\delta)\ge\varepsilon)\le\frac{2\tau}{q}+\frac{1}{q}\int_\tau^q \mathbb{P}(I(t,\delta)\ge\varepsilon)\mathrm{d}t, \quad t\ge\tau.$$

With (2.10) in hand, the uniqueness of invariant measure and its mixing property can be proved by following the steps 2 and 3 of the proof for Theorem 1.5.

Remark 2.6 Under dissipative conditions, by the Arzelà–Ascoli tightness characterization, Es-Sarhir et al. [54] and Kinnally-Williams [79] exploited existence of invariant measures for FSDEs with super-linear drift terms and positivity constraints, respectively. Applying the Itô formula, they gave the uniform boundedness for higher moments of the segment processes, which plays a key role in analyzing the diffusion terms by the Kolmogrov tightness criterion. However, (2.4) need not satisfy any dissipative conditions, and therefore the techniques adopted in [54, 79] no longer work. In this section, adopting the variation-of-constants formula and utilizing the Arzelà–Ascoli tightness characterization, we provide verifiable criterion to capture a unique invariant measure for a class of semi-linear FSDEs.

Remark 2.7 Under dissipative conditions, Itô-Nisio [72] discussed existence of stationary solutions for FSDEs. According to [72, Theorem 3], we deduce from (2.9) that (2.4), without satisfying a dissipative condition, has a stationary solution. A solution $(X(t))_{t\ge-\tau}$ of (2.4) is strong stationary, or simply stationary, if the finite-dimensional distributions are invariant under time translation, i.e.,

$$\mathbb{P}\{X(t+t_k)\in\Gamma_k,\ k=1,\dots,n\}=\mathbb{P}\{X(t_k)\in\Gamma_k,\ k=1,\dots,n\}$$

for all $t\ge 0$, $t_k\ge-\tau$ and $\Gamma_k\in\mathscr{B}(\mathbb{R}^n)$. For stationary solutions of retarded O–U processes in Hilbert spaces, we refer to, for instance, [96]. It is worth pointing out that the stationary solutions discussed for example, in [72, 96] are related to the solution process $(X(t))_{t\ge-\tau}$. Whereas the invariant measure in our case involves the segment process $(X_t)_{t\ge 0}$. Before concluding this section, we give examples to show the validity of Theorem 2.5.

Example 2.8 Consider a semi-linear SDDE

$$\mathrm{d}X(t) = -X(t-1)\mathrm{d}t + \sigma(X(t-1))\mathrm{d}W(t), \quad X_0 = \xi \in \mathscr{C}. \tag{2.19}$$

In this case, $\tau = -1$ and $\mu(\cdot) = -\delta_{-1}(\cdot)$. It is readily seen that the corresponding characteristic equation is

$$\lambda + e^{-\lambda} = 0. \tag{2.20}$$

A simple calculation using MATLAB yields that the unique root of (2.20) is

$$\lambda \approx -0.3181 + 1.3372\mathrm{i}.$$

So, one has $v_\mu \approx -0.3181$. Thus, by Theorem 2.5, we deduce that (2.19) possesses a unique invariant measure $\pi \in \mathscr{P}(\mathscr{C})$ whenever the Lipschitz constant L of σ such that $c_\lambda^2 L^2(1+e^{2\lambda}) < \lambda$ for $\lambda \in (0, 0.3181)$. Note that (2.19) does not satisfy a dissipative condition even though the Lipschitz constant of σ is sufficiently small since it is impossible to choose constants $\lambda_1 > \lambda_2 > 0$ such that

$$-2(x_1 - x_2)(y_1 - y_2) + |\sigma(y_1) - \sigma(y_2)|^2 \le -\lambda_1|x_1 - x_2|^2 + \lambda_2|y_1 - y_2|^2$$

holds for any $x_1, x_2, y_1, y_2 \in \mathbb{R}$.

Example 2.9 Consider a semi-linear SDDE

$$\mathrm{d}X(t) = \{aX(t) + bX(t-1)\}\mathrm{d}t + \sigma(X(t-1))\mathrm{d}W(t), \quad X_0 = \xi \in \mathscr{C}, \tag{2.21}$$

where $a < 0$, $b \in \mathbb{R}$ and $\sigma : \mathbb{R} \mapsto \mathbb{R}$ is Lipschitz. For this case, $\mu(\cdot) = a\delta_0(\cdot) + b\delta_{-1}(\cdot)$. Note that the associated characteristic equation is

$$\lambda - a - be^{-\lambda} = 0. \tag{2.22}$$

By [90, Theorem 1], all the roots of (2.22) have negative real parts if and only if

$$a < b < -a. \tag{2.23}$$

Then, from Theorem 2.5, (2.21) admits a unique invariant measure $\pi \in \mathscr{P}(\mathscr{C})$ provided that the Lipschitz constant L of σ is sufficiently small.

2.3 Ergodicity for FSDEs of Neutral Type

In this section, we proceed to generalize Theorem 2.5 to an FSDE of neutral type in the form

$$\mathrm{d}\Big(X(t)-\int_{[-\tau,0]}\rho(\mathrm{d}\theta)X(t+\theta)\Big)=\Big(\int_{[-\tau,0]}\mu(\mathrm{d}\theta)X(t+\theta)\Big)\mathrm{d}t+\sigma\mathrm{d}W(t) \quad (2.24)$$

with the initial value $X_0=\xi\in\mathscr{C}$, where ρ, μ are $\mathbb{R}^n\otimes\mathbb{R}^n$-valued finite signed measure on $[-\tau,0]$, $\sigma\in\mathbb{R}^n\otimes\mathbb{R}^m$.

To begin, we give an overview of the variation-of-constants formula for linear functional differential equations of neutral type. By [65, Theorem 1.1, p. 256], the following linear functional differential equation of neutral type

$$\mathrm{d}\Big(Y(t)-\int_{[-\tau,0]}\rho(\mathrm{d}\theta)Y(t+\theta)\Big)=\Big(\int_{[-\tau,0]}\mu(\mathrm{d}\theta)Y(t+\theta)\Big)\mathrm{d}t \quad (2.25)$$

with the initial data $\xi\in\mathscr{C}$ has a unique solution $(Y(t;\xi))_{t\geq-\tau}$. Let

$$v_{\rho,\mu}=\sup\{\mathrm{Re}(\lambda):\lambda\in\mathbb{C},\ \Delta_{\rho,\mu}(\lambda)=0\},$$

where

$$\Delta_{\rho,\mu}(\lambda):=\det\Big(\lambda\Big(I_{n\times n}-\int_{[-\tau,0]}e^{\lambda\theta}\rho(\mathrm{d}\theta)\Big)-\int_{[-\tau,0]}e^{\lambda\theta}\mu(\mathrm{d}\theta)\Big),\quad\lambda\in\mathbb{C}.$$

The fundamental solution or resolvent of (2.25) is the unique continuous function $\Gamma:[0,\infty)\mapsto\mathbb{R}^n\otimes\mathbb{R}^n$ satisfying

$$\mathrm{d}\Big(\Gamma(t)-\int_{[-\tau,0]}\rho(\mathrm{d}\theta)\Gamma(t+\theta)\Big)=\Big(\int_{[-\tau,0]}\mu(\mathrm{d}\theta)\Gamma(t+\theta)\Big)\mathrm{d}t$$

with the initial data $\Gamma(0)=I_{n\times n}$ and $\Gamma(\theta)=\mathbf{0}_{n\times n}$ for $\theta\in[-\tau,0)$. It plays a role analogous to the fundamental system in linear ordinary differential equations and the Green function in partial differential equations. If $v_{\rho,\mu}<0$, according to Theorem B.4, there exists $\lambda\in(0,-v_{\rho,\mu})$ such that

$$\|\Gamma(t)\|\leq c_\lambda e^{-\lambda t} \quad (2.26)$$

for some constant $c_\lambda>0$. By (B.6) and the variation-of-constants formula [6, Theorem 1], the solution of (2.24) can be represented explicitly as

$$\begin{aligned} X(t;\xi) &= Y(t;\xi) + \int_0^t \Gamma(t-s)\sigma \mathrm{d}W(s) \\ &= \Gamma(t)\xi(0) - \int_{[-\tau,0]} \rho(\mathrm{d}\theta)\Gamma(t+\theta)\xi(0) \\ &\quad + \int_{[-\tau,0]} \mu(\mathrm{d}\theta) \int_\theta^0 \Gamma(t+\theta-s)\xi(s)\mathrm{d}s \\ &\quad + \int_{[-\tau,0]} \rho(\mathrm{d}\theta) \int_\theta^0 \Gamma(t-s+\theta)\xi'(s)\mathrm{d}s + \int_0^t \Gamma(t-s)\sigma \mathrm{d}W(s) \end{aligned} \tag{2.27}$$

for any $\xi \in W^{1,2}([-\tau,0];\mathbb{R}^n)$ (the Sobolev space consisting of functions $f : [-\tau,0] \mapsto \mathbb{R}^n$ such that $f(\cdot)$ and its distributional derivative $f'(\cdot)$ belong to $L^2([-\tau,0];\mathbb{R}^n)$). For more properties of Γ, refer to Theorem B.4.

Remark 2.10 For more details on the variation-of-constants formula of linear functional differential equations of neutral type, we refer the readers to, e.g., [6, 46, 65] for finite-dimensional case, and [95, 111, 145] for infinite-dimensional setting.

Let $|||\rho||| = \sup_{1\le k\le n}\sqrt{\sum_{1\le j\le n}\|\rho_{kj}\|^2_{\mathrm{var}}}$, where $\|\rho_{kj}\|_{\mathrm{var}}$ is the total variation of ρ_{kj}.

Theorem 2.11 *Let* $v_{\rho,\mu} < 0$ *and* $|||\rho||| < 1/2$. *Then* (2.24) *has a unique invariant measure* $\pi \in \mathscr{P}(\mathscr{C})$, *which is exponentially mixing.*

Proof In view of the argument in the proof of Theorem 2.5, to complete the proof of Theorem 2.1.1, it is sufficient to show that

$$\sup_{t\ge 0}\mathbb{E}\|X_t(\xi)\|^p_\infty < \infty, \quad p > 2.$$

and that

$$\mathbb{E}\|X_t(\xi) - X_t(\eta)\|^2_\infty \lesssim e^{-\alpha t}$$

for some $\alpha > 0$.

There exists a function $\phi : \mathbb{R} \mapsto \mathbb{R}$ such that

- $\phi(x) \ge 0$ for all $x \in \mathbb{R}$;
- $\phi \in C_0^\infty(\mathbb{R})$;
- $\int_{\mathbb{R}} \phi(x)\mathrm{d}x = 1$,

where $C_0^\infty(\mathbb{R})$ denotes the collection of C^∞ functions $f : \mathbb{R} \mapsto \mathbb{R}$ with compact support; see, for instance, [56]. For any $\delta > 0$, let $\phi_\delta(x) = \delta^{-k}\phi(\frac{x}{k})$ be the standard mollifier. Let ξ_δ be the convolution of ξ and ϕ_δ, i.e.,

$$\xi_\delta(\theta) = (\xi * \phi_\delta)(\theta) = \int_{\mathbb{R}} \xi(\theta - s)\phi_\delta(s)\mathrm{d}s = \int_{\mathbb{R}} \xi(s)\phi_\delta(\theta - s)\mathrm{d}s,$$

where we set $\xi(\theta) = \mathbf{0}_n$ for $\theta \in (-\infty, -\tau) \cup (0, \infty)$. It is easy to see that $\xi_\delta \in C_0^\infty(\mathbb{R})$. Moreover, $\xi_\delta \to \xi$ uniformly on compact subsets of $\mathbb{R}$. So ξ is mollified by convolution with ϕ_δ. Observe that

$$\mathbb{E}\|X_t(\xi)\|_\infty^2 \lesssim \mathbb{E}\|X_t(\xi) - X_t(\xi_\delta)\|_\infty^2 + \mathbb{E}\|X_t(\xi_\delta)\|_\infty^2,$$

and that

$$\begin{aligned}\mathbb{E}\|X_t(\xi) - X_t(\eta)\|_\infty^2 \lesssim\, & \mathbb{E}\|X_t(\xi) - X_t(\xi_\delta)\|_\infty^2 + \mathbb{E}\|X_t(\eta) - X_t(\eta_\delta)\|_\infty^2 \\ & + \mathbb{E}\|X_t(\xi_\delta) - X_t(\eta_\delta)\|_\infty^2.\end{aligned}$$

By a straightforward calculation, one has

$$\lim_{\delta \to 0} \mathbb{E}\|X_t(\xi) - X_t(\xi_\delta)\|_\infty^2 = \lim_{\delta \to 0} \mathbb{E}\|X_t(\eta) - X_t(\eta_\delta)\|_\infty^2 = 0.$$

So, to finish the proof we need only show that

$$\sup_{t \geq 0} \mathbb{E}\|X_t(\xi_\delta)\|_\infty^p < \infty, \quad p > 2. \tag{2.28}$$

and

$$\mathbb{E}\|X_t(\xi_\delta) - X_t(\eta_\delta)\|_\infty^2 \lesssim e^{-\alpha t} \tag{2.29}$$

for some $\alpha > 0$. Observe that $\xi_\delta, \eta_\delta \in C_0^\infty(\mathbb{R})$. With (2.27) in hand, (2.29) can be done by carrying out the arguments to derive (2.10), while (2.28) can be finished by following the argument to deduce (2.9).

Remark 2.12 In fact, Theorem 2.11 can be generalized to the case of FSDEs of neutral type with multiplicative noises by following the main line of argument for Theorem 2.5. However, to avoid complicated calculations, in this section we focus only on the additive noise case.

Example 2.13 Consider an SDDE of neutral type

$$\mathrm{d}\Big(X(t) + \frac{1}{3}X(t-1)\Big) = -X(t-1)\mathrm{d}t + \sigma \mathrm{d}W(t), \quad X_0 = \xi, \tag{2.30}$$

where $a \in \mathbb{R}$ and $(W(t))_{t \geq 0}$ is a real-valued Brownian motion defined on the filtered probability space $(\Omega, \mathscr{F}, (\mathscr{F}_t)_{t \geq 0}, \mathbb{P})$. The characteristic equation associated with the deterministic counterpart of (2.30) is

$$\lambda + \Big(1 + \frac{\lambda}{3}\Big)e^{-\lambda} = 0, \quad \lambda \in \mathbb{C}. \tag{2.31}$$

Using MATLAB, we can obtain that the unique root of (2.31) is $\lambda \approx -2.313474269$. Then, by Theorem 2.1.1 we deduce that (2.30) has a unique invariant measure, which is exponentially mixing.

2.4 Ergodicity for FSPDEs Driven by Cylindrical Wiener Processes

In this section, we are concerned with ergodicity for a range of FSPDEs. First, we introduce some notation. Let $(H, \langle\cdot,\cdot\rangle_H, \|\cdot\|_H)$ and $(K, \langle\cdot,\cdot\rangle_K, \|\cdot\|_K)$ be real separable Hilbert spaces. Let $Q \in \mathscr{L}(K)$ be symmetric and nonnegative so that there exists a nonnegative and symmetric element $Q^{\frac{1}{2}} \in \mathscr{L}(K)$ such that $Q^{\frac{1}{2}} \circ Q^{\frac{1}{2}} = Q$ (see, e.g., [117, Theorem VI.9]). Let the trace of Q be finite, i.e., $\mathrm{trace}(Q) < \infty$. Then, $Q^{\frac{1}{2}} \in \mathscr{L}_{HS}(K)$ due to $\|Q^{\frac{1}{2}}\|_{HS}^2 = \mathrm{trace}(Q)$, and $S \circ Q^{\frac{1}{2}} \in \mathscr{L}_{HS}(K, H)$ for any $S \in \mathscr{L}(K, H)$. Let $K_0 = Q^{\frac{1}{2}}(K)$, which is a separable Hilbert space with the inner product given by $\langle u_0, v_0\rangle_0 := \langle Q^{-\frac{1}{2}}u_0, Q^{-\frac{1}{2}}v_0\rangle_K$ for $u_0, v_0 \in K$ (see, e.g., [113, Proposition C.0.3, p. 145]), where $Q^{-\frac{1}{2}}$ is the pseudoinverse of $Q^{\frac{1}{2}}$ in the case that Q is not one-to-one. Let $L_2^0 = \mathscr{L}_{HS}(K_0, H)$. According to [113, Proposition C.0.3 (ii), p. 145], $(Q^{\frac{1}{2}}e_k)_{k\geq 1}$ is an orthonormal basis of $(K_0, \langle\cdot,\cdot\rangle_0, \|\cdot\|_0)$ if $(e_k)_{k\geq 1}$ is an orthonormal basis of $(\mathrm{Ker}Q^{\frac{1}{2}})^{\perp}$. Thus, $\|S\|_{L_2^0} = \|S \circ Q^{\frac{1}{2}}\|_{HS}$ for each $S \in L_2^0$.

In this section, we focus on an FSPDE on $(H, \langle\cdot,\cdot\rangle_H, \|\cdot\|_H)$ in the form

$$\mathrm{d}X(t) = \{AX(t) + \Phi(X_t)\}\mathrm{d}t + \sigma \mathrm{d}W(t), \quad t > 0, \quad X_0 = \xi \in \mathscr{C}. \tag{2.32}$$

Herein, $(A, \mathscr{D}(A))$ is a self-adjoint operator on H generating a compact C_0-semigroup $(e^{tA})_{t\geq 0}$ such that $\|e^{tA}\| \leq e^{-\alpha t}$ for some $\alpha > 0$, the linear operator $\Phi : \mathscr{C} \mapsto H$ is given by

$$\Phi(\phi) := \int_{[-\tau,0]} (\mu(\mathrm{d}\theta))\phi(\theta), \quad \phi \in \mathscr{C},$$

where $\mu : [-\tau, 0] \mapsto \mathscr{L}(H)$ is of bounded variation, $\sigma \in L_2^0$, and $(W(t))_{t\geq 0}$ is a K-valued Q-Wiener process under $(\Omega, \mathscr{F}, (\mathscr{F}_t)_{t\geq 0}, \mathbb{P})$. By [11, Theorem A.1], (2.32) has a unique mild solution $(X(t))_{t\geq -\tau}$ satisfying

$$X(t) = e^{tA}\xi(0) + \int_0^t e^{(t-s)A}\Phi(X_s)\mathrm{d}s + \int_0^t e^{(t-s)A}\sigma \mathrm{d}W(s), \quad t > 0. \tag{2.33}$$

Consider the following deterministic evolution equation

$$\mathrm{d}Y(t) = \{AY(t) + \Phi(Y_t)\}\mathrm{d}t, \quad t > 0, \quad Y_0 = \xi. \tag{2.34}$$

According to [145, Theorem 1.1, p. 37], equation (2.34) has a unique mild solution $(Y(t;\xi))_{t\geq-\tau}$ with the initial data $Y_0=\xi$ such that

$$Y(t;\xi)=e^{tA}\xi(0)+\int_0^t e^{(t-s)A}\Phi(Y_s(\xi))\mathrm{d}s,\quad t>0 \tag{2.35}$$

holds. Let $(\Gamma(t))_{t\geq0}$ solve the $\mathscr{L}(H)$-valued evolution equation

$$\mathrm{d}\Gamma(t)=\{A\Gamma(t)+\Phi(\Gamma_t)\}\mathrm{d}t,\quad t>0 \tag{2.36}$$

with the initial value $\Gamma(0)=I$, the identical operator on H, and $\Gamma(\theta)=O$ for $\theta\in[-\tau,0)$, where O stands for the null operator on H. By the Laplace transform approach, the mild solution $Y(t;\xi)$ to (2.34) can be written explicitly as

$$Y(t;\xi)=\Gamma(t)\xi(0)+\int_0^t\Gamma(t-s)\Phi(\overrightarrow{\xi}_s)\mathrm{d}s,\quad t\geq0,$$

where $\overrightarrow{\xi}:[-\tau,\infty)\mapsto H$ is the right extension of ξ defined by

$$\overrightarrow{\xi}(\theta)=\xi(\theta),\quad \theta\in[-\tau,0]\quad\text{and}\quad \overrightarrow{\xi}(\theta)=0,\ \ \theta\in(0,\infty);$$

see, for instance, [93, Theorem 3.2] & [111, Theorem 2.2]. Set

$$\alpha_0:=\sup\{\mathrm{Re}(\lambda):\ \lambda\in\mathbb{C},\text{ there exists an }x\in\mathscr{D}(A)\setminus\{0\}\text{ such that }(\lambda I-A-\Phi(e^{\lambda\cdot}))x=0\},$$

where $(e^{\lambda\cdot}x)(\theta)=e^{\lambda\theta}x,\theta\in[-\tau,0]$.

Theorem 2.14 *Assume that $(A,\mathscr{D}(A))$ is a self-adjoint operator on H generating a compact C_0-semigroup $(e^{tA})_{t\geq0}$ such that $\|e^{tA}\|\leq e^{-\alpha t}$ for some $\alpha>0$ and $\alpha_0<0$, and suppose further trace$(Q)<\infty$ and $\sigma\in L_2^0$. Then, (2.32) admits a unique invariant measure π, which is exponentially mixing.*

Proof To show existence of an invariant measure, we adopt the stability-in-distribution approach (see, e.g., [14] & [155]). To complete the proof, it is sufficient to verify that

(P1) $\sup_{t\geq0}\mathbb{E}\|X_t(\xi)\|_\infty^2<\infty$ for any $\xi\in U$;
(P2) $\lim_{t\to\infty}\mathbb{E}\|X_t(\xi)-X_t(\eta)\|_\infty^2=0$ for any $\xi,\eta\in U$,

in which U is a bounded subset of $\mathscr{C}$. In what follows, we intend to show that (P1) and (P2) are fulfilled, respectively, for (2.32). By [93, Proposition 4.1], the unique mild solution $X(t;\xi)$ with the initial value $X_0=\xi$ can be represented explicitly by

$$
\begin{aligned}
X(t;\xi) &= Y(t;\xi) + \int_0^t \Gamma(t-s)\sigma \mathrm{d}W(s) \\
&= \Gamma(t)\xi(0) + \int_0^t \Gamma(t-s)\Phi(\overrightarrow{\xi}_s)\mathrm{d}s + \int_0^t \Gamma(t-s)\sigma \mathrm{d}W(s). \qquad (2.37)
\end{aligned}
$$

Moreover, for any $\lambda \in (0, -\alpha_0)$, by virtue of [94, Lemma 2.1] there exists $c_\lambda > 0$ such that

$$
\|\Gamma(t)\| \le c_\lambda e^{-\lambda t}. \qquad (2.38)
$$

This, together with (2.37) and Itô's isometry, gives that

$$
\begin{aligned}
\mathbb{E}\|X(t;\xi)\|_H^2 &\lesssim \|\xi\|_\infty^2 + \int_0^t \mathrm{trace}(\Gamma(s)\sigma Q\sigma^*\Gamma(s)^*)\mathrm{d}s \\
&\lesssim \|\xi\|_\infty^2 + \int_0^t \|\Gamma(s)\sigma Q^{\frac{1}{2}}\|_{HS}^2 \mathrm{d}s \qquad (2.39) \\
&\lesssim \|\xi\|_\infty^2 + \|\sigma\|_{L_2^0}^2 \int_0^t \|\Gamma(s)\|\mathrm{d}s \\
&\lesssim 1 + \|\xi\|_\infty^2.
\end{aligned}
$$

Recall that $\|e^{tA}\| \le e^{-\alpha t}$ for some $\alpha > 0$. Next, for any $t \ge \tau$ observe from (2.33) and (2.39) that

$$
\begin{aligned}
&\mathbb{E}\|X(t;\xi)\|_\infty^2 \\
&\lesssim \mathbb{E}\|X(t-\tau;\xi)\|_H^2 + \mathbb{E}\Big(\sup_{t-\tau\le s\le t}\Big\|\int_{t-\tau}^s e^{(s-u)A}\Phi(X_u)\mathrm{d}u\Big\|_H^2\Big) \\
&\quad + \mathbb{E}\Big(\sup_{t-\tau\le s\le t}\Big\|\int_{t-\tau}^s e^{(s-u)A}\sigma \mathrm{d}W(u)\Big\|_H^2\Big) \qquad (2.40) \\
&\lesssim 1 + \int_{t-\tau}^t \mathbb{E}\|\Phi(X_u)\|_H^2 \mathrm{d}u + \Big(\mathbb{E}\Big(\sup_{t-\tau\le s\le t}\Big\|\int_{t-\tau}^s e^{(s-u)A}\sigma \mathrm{d}W(u)\Big\|_H^{2r}\Big)\Big)^{1/r} \\
&\lesssim 1 + \|\xi\|_\infty^2,
\end{aligned}
$$

where in the second step we have utilized the Hölder inequality with $r > 1$ for the time integral and in the last step used [36, Proposition 7.3, p. 184] for the stochastic convolution. Hence, (P1) follows. Hereinafter, we intend to verify (P2). From (2.37), one has

$$
X(t;\xi)-X(t;\eta) = \Gamma(t)(\xi(0)-\eta(0))+\int_0^t \Gamma(t-s)(\Phi(\overrightarrow{\xi}_s)-\Phi(\overrightarrow{\eta}_s))\mathrm{d}s. \qquad (2.41)
$$

Moreover, for any $t \ge \tau$ by the notions of $\overrightarrow{\xi}$ and $\overrightarrow{\eta}$, we derive from Hölder's inequality that

$$
\begin{aligned}
&\Big\|\int_0^t \Gamma(t-s)(\Phi(\overrightarrow{\xi}_s)-\Phi(\overrightarrow{\eta}_s))\mathrm{d}s\Big\|_H^2\\
&\lesssim \Big\|\int_0^\tau \Gamma(t-s)(\Phi(\overrightarrow{\xi}_s)-\Phi(\overrightarrow{\eta}_s))\mathrm{d}s\Big\|_H^2+\Big\|\int_\tau^t \Gamma(t-s)(\Phi(\overrightarrow{\xi}_s)-\Phi(\overrightarrow{\eta}_s))\mathrm{d}s\Big\|_H^2\\
&\lesssim \int_0^\tau \|\Gamma(t-s)\|\cdot\|\Phi(\overrightarrow{\xi}_s)-\Phi(\overrightarrow{\eta}_s)\|_H^2\mathrm{d}s\\
&\lesssim e^{-\lambda t}\|\xi-\eta\|_\infty^2,
\end{aligned}
$$

which, combining (2.38) with (2.41), yields that

$$\|X(t;\xi)-X(t;\eta)\|_H^2 \lesssim e^{-\lambda t}\|\xi-\eta\|_\infty^2, \quad t\geq -\tau.$$

So one has

$$\|X(t+\theta;\xi)-X(t+\theta;\eta)\|_H^2 \lesssim e^{-\lambda(t+\theta)}\|\xi-\eta\|_\infty^2 \lesssim e^{-\lambda t}\|\xi-\eta\|_\infty^2, \quad \theta\in[-\tau,0].$$

Thus we arrive at

$$\mathbb{E}\|X(t;\xi)-X(t;\eta)\|_\infty^2 \lesssim e^{-\lambda t}\|\xi-\eta\|_\infty^2. \tag{2.42}$$

Consequently, (P2) holds.

Example 2.15 Consider the following SPDE with memory

$$
\begin{cases}
\frac{\partial}{\partial t}u(t,x)=\Delta u(t,x)+bu(t-\tau,x)+c(x)\dot{W}(t), & t>0, \ x\in[0,\pi],\\
u(t,0)=u(t,\pi)=0, & t\geq 0,\\
u(t,x)=\phi(t,x), & t\in[-\tau,0], \ x\in[0,\pi],
\end{cases}
\tag{2.43}
$$

where $b\in\mathbb{R}$, $c(\cdot)$, $\phi(t,\cdot)\in L^2([0,\pi])$, and $(W(t))_{t\geq 0}$ is a scalar Brownian motion. Let $A=\Delta$ be the Laplacian. Note that the characteristic equation associated with the deterministic counterpart of (2.43) is

$$(\lambda I - A - be^{-\lambda\tau}I)x=0, \quad x\in\mathscr{D}(A)\setminus\{0\}. \tag{2.44}$$

Since the eigenvalues of A is $-n^2$, the eigenvalue of (2.44) satisfies

$$\lambda+n^2-be^{-\lambda\tau}=0, \quad n=1,2\ldots. \tag{2.45}$$

By [90, Theorem 1], all the roots of (2.45) with each fixed $n\geq 1$ have negative real parts if and only if

$$-n^2<b<n^2.$$

So (2.43) admits a unique invariant measure whenever $-1<b<1$.

In what follows, we consider the case $Q=I$ so that $\mathrm{trace}(Q)=\infty$, i.e., $(W(t))_{t\geq 0}$ is a cylindrical Wiener process.

Theorem 2.16 *Assume that* $(A, \mathscr{D}(A))$ *is a self-adjoint operator on* H *generating a compact* C_0*-semigroup* $(e^{tA})_{t\geq 0}$ *such that* $\|e^{tA}\| \leq e^{-\alpha t}$ *for some* $\alpha > 0$ *and* $\alpha_0 < 0$*. Suppose further that*

$$\sup_{t\geq 0}\left(\int_0^t s^{-2\alpha}\|e^{sA}\|_{HS}^2 ds\right) < \infty, \tag{2.46}$$

and

$$\sup_{t\geq 0}\left(\int_0^t trace(\Gamma(s)\sigma Q\sigma^*\Gamma(s)^*)ds\right) < \infty. \tag{2.47}$$

Then, (2.32) *admits a unique invariant measure* π*, which is exponentially mixing.*

Proof Theorem 2.16 can be proved by a slight modification of the argument for Theorem 2.14. To complete the proof, we still need to verify (P1) and (P2), respectively. Due to (2.47), we obtain from the first display of (2.39) that

$$\mathbb{E}\|X(t;\xi)\|_H^2 \lesssim 1 + \|\xi\|_\infty^2. \tag{2.48}$$

From the second inequality of (2.40), we deduce from (2.48) and [36, Proposition 7.9, p. 196] that

$$\begin{aligned}\mathbb{E}\|X(t;\xi)\|_\infty^2 &\lesssim 1 + \|\xi\|_\infty^2 + \int_{t-\tau}^t s^{-2\alpha}\|e^{sA}\|_{HS}^2 \mathrm{d}s\\ &\lesssim 1 + \|\xi\|_\infty^2.\end{aligned}$$

As a result, (P1) follows. Also, (P2) holds as (2.42) shows.

2.5 Ergodicity for FSDEs Driven by Jump Processes

In the previous sections, we obtained existence and uniqueness of invariant measures for functional stochastic equations with continuous sample paths. In this section, we consider the case of FSDEs driven by jump processes. The discontinuity of the sample path makes the problem more difficult to deal with. Although the variation-of-constants approach can still be used, the verification of the tightness cannot be done as in the case of continuous sample paths. To overcome the difficulty, we use the Kurtz tightness criterion (see Lemma 2.17 below) to treat the underlying problem.

We start with some terminologies and notation. Let $(Z(t))_{t\geq 0}$ be an n-dimensional Lévy process defined on the stochastic basis $(\Omega, \mathscr{F}, (\mathscr{F}_t)_{t\geq 0}, \mathbb{P})$. For each $t \geq 0$, $Z(t)$ is infinitely divisible by virtue of [3, Proposition 1.3.1, p. 40]. By the Lévy–Khintchine formula (see, e.g., [3, Theorem 1.2.14, p. 28]), $\Psi(\xi)$, the symbol or characteristic exponent of $Z(t)$, admits the form

$$\Psi(\xi) = i\langle b, \xi\rangle - \frac{1}{2}\langle \xi, A\xi\rangle + \int_{\mathbb{R}^m-\{0\}} \{e^{i\langle \xi, z\rangle} - 1 - i\langle \xi, z\rangle \mathbf{1}_{\{|z|<1\}}\}\pi(\mathrm{d}z),$$

in which $b \in \mathbb{R}^n$, $A \in \mathbb{R}^n \otimes \mathbb{R}^n$ is positive definite and symmetric, and $\pi(\cdot)$ is the Lévy measure, i.e., a σ-finite measure on $\mathbb{R}^n - \{0\}$ such that

$$\int_{\mathbb{R}^n-\{0\}} (1 \wedge |z|^2)\pi(\mathrm{d}z) < \infty.$$

The lemma below is concerned with Kurtz's criterion on tightness of laws on $\mathscr{D}$.

Lemma 2.17 ([55, Theorem 8.6, p. 137–138]) *For each* $t \geq 0$, *let* $Y^t(\cdot) \in D([0, \tau]; \mathbb{R}^n)$ *and assume that*

$$\lim_{K_1\to\infty} \limsup_{t\to\infty} \mathbb{P}\Big(\sup_{0\leq s\leq T} |Y^t(s)| \geq K_1\Big) = 0 \textit{ for each } T \leq \tau,$$

and, for $0 \leq u \leq \delta$ *and* $v \leq T$,

$$\mathbb{E}_v^t\{|Y^t(u+v) - Y^t(u)| \wedge 1\} \leq \mathbb{E}_v^t \gamma(t, \delta)$$
$$\lim_{\delta\to 0} \limsup_{t\to\infty} \mathbb{E}\gamma(t, \delta) = 0,$$

where $\mathbb{E}_v^t$ *denotes the conditional expectation with respect to* $\mathscr{F}_v^t$, *the minimal* σ*-algebra that measures* $(Y^t(s))_{0\leq s\leq v}$. *Then* $(\mathscr{L}(Y^t(s)), s \in [0, \tau])_{t\geq 0}$ *is tight in* $D([0, \tau]; \mathbb{R}^n)$.

Consider a retarded O–U process driven by a Lévy process with the Lévy triple $(0, 0, \pi)$ in the form

$$\mathrm{d}X(t) = \Big(\int_{[-\tau,0]} \mu(\mathrm{d}\theta)X(t+\theta)\Big)\mathrm{d}t + \sigma \mathrm{d}Z(t), \quad X_0 = \xi \in \mathscr{D}, \tag{2.49}$$

where $\mu(\cdot)$ is an $\mathbb{R}^n \otimes \mathbb{R}^n$-valued finite signed measure on $[-\tau, 0]$, and $\sigma \in \mathbb{R}^n \otimes \mathbb{R}^n$. Hereinafter, $v_\mu \in \mathbb{R}$ is defined as in (2.5). The main result in this section is presented as follows:

Theorem 2.18 *Let* $v_\mu < 0$ *and assume further that*

$$\int_{|z|>1} |z|\pi(dz) < \infty. \tag{2.50}$$

Then, (2.49) *has a unique invariant measure* $\mu \in \mathscr{P}(\mathscr{D})$.

Proof For each integer $q \geq 1$, set

$$\mu_q(\cdot) := \frac{1}{q}\int_0^q \mathbb{P}(t, \xi, \cdot)\mathrm{d}t,$$

where $\mathbb{P}(t,\xi,\cdot)$ is the Markovian transition kernel of $X_t(\xi)$. If $(\mathscr{L}(X_t(\xi)))_{t\geq\tau}$ is tight under the Skorohod metric d_S, for any $\varepsilon > 0$ there exists a compact subset $U \in \mathscr{B}(\mathscr{D})$ such that

$$\mathbb{P}(X_t(\xi) \in U) \leq 1-\varepsilon \text{ and hence } \mu_q(U) \leq 1-\varepsilon.$$

Thus $(\mu_q(\cdot))_{q\geq 1}$ is tight. Recall that $(X_t(\xi))_{t\geq 0}$ is Markovian by Theorem B.2 (see also [119, Proposition 3.3]) and eventually Feller due to [119, Proposition 3.5], i.e., for all $t \geq \tau$, P_t maps $C_b(\mathscr{D})$ into itself. As a consequence, the Krylov–Bogoliubov theorem (see, e.g., [37, Theorem 3.1.1, p. 21]) implies that (2.49) has an invariant measure $\pi \in \mathscr{P}(\mathscr{D})$.

For $t > 0$ and $\Gamma \in \mathscr{B}(\mathbb{R}^n - \{0\})$, define the Poisson random measure generated by $Z(t)$ by

$$N(t,\Gamma) = \sum_{s\in(0,t]} \mathbf{1}_\Gamma(\Delta Z(s)),$$

where $\Delta Z(t) := Z(t) - Z(t-)$ for any $t \geq 0$, and the compensated Poisson random measure by

$$\widetilde{N}(t,\Gamma) = N(t,\Gamma) - t\pi(\Gamma).$$

By the Lévy-Itô decomposition (see, e.g., [3,Theorem 2.4.16, p. 108]), one gets

$$Z(t) = \int_{|z|\leq 1} z\widetilde{N}(t,\mathrm{d}z) + \int_{|z|>1} zN(t,\mathrm{d}z). \tag{2.51}$$

By the variation-of-constants formula (see, e.g., Gushchin–Küchler [60]), the solution of (2.49) can be written explicitly as

$$\begin{aligned} X(t;\xi) = \Gamma(t)\xi(0) &+ \int_{[-\tau,0]} \mu(\mathrm{d}\theta) \int_\theta^0 \Gamma(t+\theta-s)\xi(s)\mathrm{d}s \\ &+ \int_0^t \Gamma(t-s)\sigma \mathrm{d}Z(s), \end{aligned} \tag{2.52}$$

where $(\Gamma(t))_{t\geq -\tau}$ is the fundamental solution of (2.6). Substituting (2.51) into (2.52) leads to

$$\begin{aligned} X(t;\xi) &= \Gamma(t)\xi(0) + \int_{[-\tau,0]} \mu(\mathrm{d}s) \int_s^0 \Gamma(t+s-u)\xi(u)\mathrm{d}u \\ &\quad + \int_0^t \int_{|z|\leq 1} \Gamma(t-s)\sigma z\widetilde{N}(\mathrm{d}s,\mathrm{d}z) + \int_0^t \int_{|z|>1} \Gamma(t-s)\sigma zN(\mathrm{d}s,\mathrm{d}z) \\ &=: \sum_{j=1}^4 I_j(t). \end{aligned}$$

By (2.8), it is easy to see that

$$\sup_{t\geq 0}\mathbb{E}(|I_1(t)|+|I_2(t)|)\lesssim \|\xi\|_\infty.$$

By virtue of the Hölder inequality, the Itô isometry, and (2.8), we have

$$\sup_{t\geq 0}\mathbb{E}|I_3(t)|\leq \beta^{1/2}\|\sigma\|\sup_{t\geq 0}\Big(\int_0^t\|\Gamma(t-s)\|^2\mathrm{d}s\Big)^{1/2}<\infty,$$

where

$$\beta:=\int_{|z|\leq 1}|z|^2\pi(\mathrm{d}z)<\infty \tag{2.53}$$

with $\pi(\cdot)$ being the Lévy measure. Also, by (2.8) it follows from (2.50) that

$$\sup_{t\geq 0}\mathbb{E}|I_4(t)|\leq \|\sigma\|\int_{|z|>1}|z|\pi(\mathrm{d}z)\sup_{t\geq 0}\Big(\int_0^t\|\Gamma(t-s)\|\mathrm{d}s\Big)<\infty.$$

Hence we arrive at

$$\sup_{t\geq 0}\mathbb{E}|X(t;\xi)|<\infty. \tag{2.54}$$

For any $t\geq\tau$, we derive from (2.51) and (2.54) that

$$\begin{aligned}\mathbb{E}\|X_t(\xi)\|_\infty &\leq \mathbb{E}|X(t-\tau;\xi)|+\int_{t-\tau}^t\mathbb{E}\Big|\int_{[-\tau,0]}\mu(\mathrm{d}\theta)X(u+\theta;\xi)\Big|\mathrm{d}u\\ &\quad+\mathbb{E}\Big(\sup_{t-\tau\leq s\leq t}\Big|\int_{t-\tau}^s\int_{|z|\leq 1}\sigma z\widetilde{N}(\mathrm{d}t,\mathrm{d}z)\Big|\Big)\\ &\quad+\mathbb{E}\Big(\sup_{t-\tau\leq s\leq t}\Big|\int_{t-\tau}^s\int_{|z|>1}\sigma zN(\mathrm{d}u,\mathrm{d}z)\Big|\Big)\\ &\leq c+\mathbb{E}\Big(\sup_{t-\tau\leq s\leq t}\Big|\int_{t-\tau}^s\int_{|z|\leq 1}\sigma z\widetilde{N}(\mathrm{d}u,\mathrm{d}z)\Big|\Big)\\ &\quad+\tau\|\sigma\|\int_{|z|>1}|z|\pi(\mathrm{d}z),\end{aligned} \tag{2.55}$$

where in the last step we have used that, by [3, Theorem 2.3.8, p. 91],

$$\begin{aligned}\mathbb{E}\Big(\sup_{t-\tau\leq s\leq t}\Big|\int_{t-\tau}^s\int_{|z|>1}\sigma zN(\mathrm{d}u,\mathrm{d}z)\Big|\Big)&\leq \|\sigma\|\mathbb{E}\int_{t-\tau}^t\int_{|z|>1}|z|N(\mathrm{d}u,\mathrm{d}z)\\ &=\|\sigma\|\tau\int_{|z|>1}|z|\pi(\mathrm{d}z).\end{aligned} \tag{2.56}$$

Next, by B–D–G's inequality [116, Theorem 48, p. 193] and Jensen's inequality, one finds that

$$\begin{aligned}\mathbb{E}\Big(\sup_{t-\tau\le s\le t}\Big|\int_{t-\tau}^{s}\int_{|z|\le 1}\sigma z\widetilde{N}(du,dz)\Big|\Big) &\lesssim \|\sigma\|\mathbb{E}\sqrt{\int_{t-\tau}^{t}\int_{|z|\le 1}|z|^2N(du,dz)}\\ &\lesssim \|\sigma\|\sqrt{\mathbb{E}\int_{t-\tau}^{t}\int_{|z|\le 1}|z|^2N(du,dz)} \\ &\lesssim \|\sigma\|\sqrt{\tau\beta},\end{aligned} \tag{2.57}$$

where in the last step we have utilized [3, Theorem 2.3.8, p. 91], and $\beta > 0$ was given in (2.53). Consequently, (2.55) and (2.57) yield that

$$\sup_{t\ge 0}\mathbb{E}\|X_t(\xi)\|_\infty < \infty. \tag{2.58}$$

Set

$$\mathbb{E}_s(\cdot) := \mathbb{E}(\cdot|\mathscr{F}_s), \quad s\ge 0.$$

For $\theta\in[-\tau,0)$ and $\widetilde{\theta}\in[0,\delta]$, where $\delta>0$ is an arbitrary constant such that $\theta+\delta\in[-\tau,0]$, (2.49) and (2.51) lead to

$$\begin{aligned}\mathbb{E}_{t+\theta}|X_t(\theta+\widetilde{\theta})-X_t(\theta)| &= \mathbb{E}_{t+\theta}|X(t+\theta+\widetilde{\theta})-X(t+\theta)|\\ &\lesssim \int_{t+\theta}^{t+\theta+\delta}\mathbb{E}_{t+\theta}\Big\{\Big|\int_{[-\tau,0]}\mu(d\theta)X(s+\theta)\\ &\quad+\int_{|z|\le 1}\sigma z\widetilde{N}(ds,dz)+\int_{|z|>1}\sigma zN(ds,dz)\Big|\Big\}.\end{aligned}$$

It follows that there is a $\gamma_0(t,\delta)$ satisfying

$$\mathbb{E}_{t+\theta}|X(t+\theta+\widetilde{\theta})-X(t+\theta)|\le \mathbb{E}_{t+\theta}\gamma_0(t,\delta).$$

Taking expectation, following the arguments to derive (2.56) and (2.58), respectively, and taking $\limsup_{t\to\infty}$ followed by $\lim_{\delta\to 0}$, we obtain from (2.54) that

$$\lim_{\delta\to 0}\limsup_{t\to\infty}\mathbb{E}\gamma_0(t,\delta)=0. \tag{2.59}$$

In view of Lemma 2.17 and the time shift $t\mapsto t-\tau$, we conclude from (2.58) and (2.59) that $(X_t(\xi))_{t\ge\tau}$ is tight under the Skorohod metric d_S. Consequently, the solution of (2.49) enjoys an invariant measure.

Next, from (2.8) and (2.52), we get that

$$
\begin{aligned}
&\mathbb{E}\|X_t(\xi) - X_t(\eta)\|_\infty \\
&= \mathbb{E}\Big(\sup_{-\tau\le\theta\le 0}\Big|\Gamma(t+\theta)(\xi(0)-\eta(0)) \\
&\quad + \int_{[-\tau,0]}\mu(\mathrm{d}u)\int_u^0 \Gamma(t+\theta+u-s)(\xi(s)-\eta(s))\mathrm{d}s\Big|\Big) \\
&\lesssim e^{-\lambda t}\|\xi-\eta\|_\infty, \quad t\ge\tau.
\end{aligned}
$$

This yields that

$$
\begin{aligned}
&\lim_{t\to\infty}\|\mathbb{P}(t,\xi,\cdot)-\mathbb{P}(t,\eta,\cdot)\|_{\mathrm{var}} \\
&= \lim_{t\to\infty}\sup_{\|\varphi\|_{\mathrm{Lip}}=1}|\mathbb{E}\varphi(X_t(\xi))-\mathbb{E}\varphi(X_t(\eta))| = 0, \qquad (2.60)
\end{aligned}
$$

where $\|\varphi\|_{\mathrm{Lip}}$ is the Lipschitz constant of φ w.r.t. the Skorohod metric d_S. If $\mu'(\cdot)\in\mathscr{P}(\mathscr{D})$ is also an invariant measure, then, by the invariance, one has

$$
\|\mu-\mu'\|_{\mathrm{var}} \le \int_{\mathscr{D}\times\mathscr{D}}\|\mathbb{P}(t,\xi,\cdot)-\mathbb{P}(t,\eta,\cdot)\|_{\mathrm{var}}\mu(\mathrm{d}\xi)\mu'(\mathrm{d}\eta). \qquad (2.61)
$$

Thus, the uniqueness of invariant measure follows by taking $t\to\infty$ in (2.61) and utilizing (2.60).

Remark 2.19 From (2.50), $\mathbb{E}|Z(t)|<\infty$ for all $t>0$ because of [3, Theorem 2.5.2, p. 132]. (2.49) incorporates retarded O–U processes driven by symmetric α-stable processes with $1<\alpha<2$, which have finite pth moment for $p\in(0,\alpha)$, and subordinate Brownian motions $W_{S(t)}$, i.e., $W(t)$ is a standard Brownian motion and $S(t)$ is an $\alpha/2$-stable subordination (i.e., a real-valued Lévy process with nondecreasing sample paths). For more details on stable distributions and subordinators, we refer to Applebaum [3, p. 33–62].

Remark 2.20 In many applications, one often encounters the so-called jump diffusion models, in which both Brownian type of noise and Lévy process appear. In view of our results in Section 2.2 and the current section, in lieu of (2.4) or (2.49), we can consider a process of the form

$$
\mathrm{d}X(t) = \Big(\int_{[-\tau,0]}\mu(\mathrm{d}\theta)X(t+\theta)\Big)\mathrm{d}t + \sigma(X_t)\mathrm{d}W(t) + \sigma_0\mathrm{d}Z(t),\ X_0=\xi\in\mathscr{D}, \qquad (2.62)
$$

where $\sigma:\mathscr{D}\mapsto\mathbb{R}^n\otimes\mathbb{R}^m$ such that (H2) in Chapter 1 holds. Continue to use the variation-of-constants formula. Comparing to the development in Theorem 2.18, we need to deal with an additional martingale term in the verification of tightness. This term can be easily handled using B–D–G's inequality. The results of Theorem 2.18 continue to hold.

Remark 2.21 Examining the proof of Theorem 2.18, the technique employed therein applies to (2.49) with the Lévy triple $(0, A, \pi)$ and an FSDE with jumps

$$dX(t) = \Big(\int_{[-\tau,0]} \mu(d\theta)X(t+\theta)\Big)dt + \sigma(X_{t-})dZ(t), \quad X_0 = \xi \in \mathscr{D}, \tag{2.63}$$

where $\sigma : \mathscr{D} \mapsto \mathbb{R}^n \otimes \mathbb{R}^n$ is uniformly bounded.

Nevertheless, if the Lévy process $Z(t)$ has a finite second moment, the uniform boundedness of σ can indeed be removed as the following theorem shows. As we mention above, $v_\mu \in \mathbb{R}$ below is defined in (2.5). We assume that there exists an $L > 0$ such that

$$\|\sigma(\phi) - \sigma(\psi)\|_{HS}^2 \le L\Big(|\phi(0) - \psi(0)|^2 + \int_{[-\tau,0]} |\phi(\theta) - \psi(\theta)|^2 \rho(d\theta)\Big) \tag{2.64}$$

for any $\xi, \eta \in \mathscr{D}$.

Theorem 2.22 *Let $v_\mu < 0$ and (2.64) hold for a sufficiently small $L > 0$, and suppose further that*

$$\vartheta := \int_{|z|\ge 1} |z|^2 \pi(dz) < \infty. \tag{2.65}$$

Then (2.63) *has a unique invariant measure $\pi \in \mathscr{P}(\mathscr{D})$.*

Proof By the variation-of-constants formula (see, e.g., [119, Theorem 3.1]), one has

$$\begin{aligned} X(t;\xi) = \Gamma(t)\xi(0) &+ \int_{[-\tau,0]} \mu(d\theta) \int_\theta^0 \Gamma(t+\theta-s)\xi(s)ds \\ &+ \int_0^t \Gamma(t-s)\sigma(X_{s-}(\xi))dZ(s), \end{aligned}$$

where $(\Gamma(t))_{t\ge -\tau}$ is the fundamental solution to (2.6). By (2.65), the Lévy-Itô decomposition (see, e.g., [3, Theorem 2.4.16, p. 126]) gives that

$$Z(t) = \Lambda t + W(t) + \int_{z\neq 0} z\widetilde{N}(t, dz), \quad \Lambda \in \mathbb{R}^n. \tag{2.66}$$

Thus, we have

$$\begin{aligned} X(t;\xi) = \Gamma(t)\xi(0) &+ \int_{[-\tau,0]} \mu(d\theta) \int_\theta^0 \Gamma(t+\theta-s)\xi(s)ds \\ &+ \int_0^t \Gamma(t-s)\sigma(X_s(\xi))\Lambda ds \\ &+ \int_0^t \Gamma(t-s)\sigma(X_s(\xi))dW(s) + \int_0^t \int_{z\neq 0} \Gamma(t-s)\sigma(X_{s-}(\xi))z\widetilde{N}(ds, dz). \end{aligned}$$

According to (2.8), there exist $\varepsilon, \delta \in (0, -\lambda_0)$ with $\varepsilon + \delta \in (0, -\lambda_0)$ such that

$$\alpha := \int_0^\infty e^{2\varepsilon t}\|\Gamma(t)\|^2 \mathrm{d}t < \infty \quad \beta := \int_0^\infty e^{2\varepsilon t}\|\dot{\Gamma}(t)\|^2 \mathrm{d}t < \infty, \tag{2.67}$$

and

$$\kappa := \int_0^\infty e^{2\delta t}\|\dot{\Gamma}_1(t)\|^2 \mathrm{d}t < \infty, \tag{2.68}$$

where $\dot{\Gamma}(t)$ denotes the derivative of $\Gamma(t)$, and $\Gamma_1(t) = e^{\varepsilon t}\Gamma(t)$. For any $\varepsilon \in (0, -\lambda_0)$ such that (2.67) holds, it is easily seen that

$$\begin{aligned}
e^{2\varepsilon t}\mathbb{E}|X(t;\xi)|^2 &\le 5\Big\{|e^{\varepsilon t}\Gamma(t)\xi(0)|^2 + \Big|\int_{[-\tau,0]}\mu(\mathrm{d}\theta)\int_\theta^0 e^{\varepsilon t}\Gamma(t+\theta-s)\xi(s)\mathrm{d}s\Big|^2 \\
&\quad + \mathbb{E}\Big|\int_0^t e^{\varepsilon t}\Gamma(t-s)\sigma(X_s(\xi))A\mathrm{d}s\Big|^2 \\
&\quad + \mathbb{E}\Big|\int_0^t e^{\varepsilon t}\Gamma(t-s)\sigma(X_s(\xi))\mathrm{d}W(s)\Big|^2 \\
&\quad + \mathbb{E}\Big|\int_0^t\int_{z\neq 0} e^{\varepsilon t}\Gamma(t-s)\sigma(X_{s-}(\xi))z\widetilde{N}(\mathrm{d}s,\mathrm{d}z)\Big|^2\Big\} \\
&=: 5\{\Upsilon_1(t)+\Upsilon_2(t)+\Upsilon_3(t)+\Upsilon_4(t)+\Upsilon_5(t)\}.
\end{aligned}$$

By Hölder's inequality, we observe from (2.67) that

$$\begin{aligned}
\Upsilon_3(t) &= \mathbb{E}\Big|\int_0^t e^{\varepsilon(t-s)}\Gamma(t-s)e^{\varepsilon s}\sigma(X_s(\xi))A\mathrm{d}s\Big|^2 \\
&\le \|A\|^2\int_0^t e^{2\varepsilon s}\|\Gamma(s)\|^2\mathrm{d}s\int_0^t e^{2\varepsilon s}\mathbb{E}\|\sigma(X_s(\xi))\|_{HS}^2\mathrm{d}s \\
&\le \alpha\|A\|^2\int_0^t e^{2\varepsilon s}\mathbb{E}\|\sigma(X_s(\xi))\|_{HS}^2\mathrm{d}s.
\end{aligned}$$

Next, by virtue of Fubini's theorem and (2.68), it follows that

$$
\begin{aligned}
\Upsilon_4(t) &= \mathbb{E}\Big|\int_0^t \Big\{I+\int_0^{t-s}\dot{\Gamma}_1(u)\mathrm{d}u\Big\}e^{\varepsilon s}\sigma(X_s(\xi))\mathrm{d}W(s)\Big|^2\\
&\le 2\mathbb{E}\Big|\int_0^t e^{\varepsilon s}\sigma(X_s(\xi))\mathrm{d}W(s)\Big|^2\\
&\quad + 2\mathbb{E}\Big|\int_0^t e^{\delta u}\dot{\Gamma}_1(u)e^{-\delta u}\int_0^{t-u} e^{\varepsilon s}\sigma(X_s(\xi))\mathrm{d}W(s)\mathrm{d}u\Big|^2\\
&\le 2\mathbb{E}\Big|\int_0^t e^{\varepsilon s}\sigma(X_s(\xi))\mathrm{d}W(s)\Big|^2 + 2\kappa\int_0^t e^{-2\delta u}\mathbb{E}\Big|\int_0^{t-u} e^{\varepsilon s}\sigma(X_s(\xi))\mathrm{d}W(s)\Big|^2\mathrm{d}u\\
&\le 2\int_0^t e^{2\varepsilon s}\mathbb{E}\|\sigma(X_s(\xi))\|_{HS}^2\mathrm{d}s + 2\kappa\int_0^t e^{-2\delta u}\int_0^{t-u} e^{2\varepsilon s}\mathbb{E}\|\sigma(X_s(\xi))\|_{HS}^2\mathrm{d}s\mathrm{d}u\\
&\le 2\int_0^t e^{2\varepsilon s}\mathbb{E}\|\sigma(X_s(\xi))\|_{HS}^2\mathrm{d}s + 2\kappa\int_0^t\int_0^{t-s} e^{-2\delta u}\mathrm{d}u e^{2\varepsilon s}\mathbb{E}\|\sigma(X_s(\xi))\|_{HS}^2\mathrm{d}s\\
&\le 2(1+\kappa\delta^{-1})\int_0^t e^{2\varepsilon s}\mathbb{E}\|\sigma(X_s(\xi))\|_{HS}^2\mathrm{d}s.
\end{aligned}
$$

Similarly, we have

$$
\Upsilon_5(t)\le 2(\beta+\vartheta)(1+\kappa\delta^{-1})\int_0^t e^{2\varepsilon s}\mathbb{E}\|\sigma(X_s(\xi))\|_{HS}^2\mathrm{d}s,
$$

where $\beta,\vartheta>0$ are introduced in (2.53) and (2.65), respectively. Furthermore, observe from (2.64) that

$$
\begin{aligned}
&\int_0^t e^{2\varepsilon s}\mathbb{E}\|\sigma(X_s(\xi))\|_{HS}^2\mathrm{d}s\\
&\le c(e^{2\varepsilon t}+\|\xi\|_\infty^2)+2(1+\varepsilon')L^2(1+e^{2\varepsilon\tau}\|\rho\|_{\mathrm{var}}^2)\int_0^t e^{2\varepsilon s}\mathbb{E}|X(s;\xi)|^2\mathrm{d}s
\end{aligned}
$$

for any $\varepsilon'>0$. Therefore, we arrive at

$$
\begin{aligned}
e^{2\varepsilon t}\mathbb{E}|X(t;\xi)|^2 \le c(e^{2\varepsilon t}+\|\xi\|_\infty^2) &+ 10\{\alpha\|A\|^2+2(1+\beta+\vartheta)(1+\kappa\delta^{-1})\}\\
&\times(1+\varepsilon')L^2(1+e^{2\varepsilon\tau}\|\rho\|_{\mathrm{var}}^2)\int_0^t e^{2\varepsilon s}\mathbb{E}|X(s;\xi)|^2\mathrm{d}s.
\end{aligned}
$$

So the Gronwall inequality leads to

$$
\sup_{t\ge 0}\mathbb{E}|X(t,\xi)|^2 \lesssim 1+\|\xi\|_\infty^2, \tag{2.69}
$$

whenever $L>0$ is sufficiently small. In the light of B–D–G's inequality (see, e.g., [116, Theorem 48, p. 193]), besides (2.64), one derives from (2.66) and (2.69) that

$$
\begin{aligned}
\mathbb{E}\|X_t(\xi)\|_\infty^2 &\lesssim 1+\mathbb{E}\Big(\sup_{t-\tau\le s\le t}\Big|\int_{t-\tau}^s \sigma(X_s)\mathrm{d}W(s)\Big|^2\Big)\\
&\quad+\mathbb{E}\Big(\sup_{t-\tau\le s\le t}\Big|\int_{t-\tau}^s\int_{z\neq 0}\sigma(X_{s-})z\widetilde{N}(\mathrm{d}s,\mathrm{d}z)\Big|^2\Big)\\
&\lesssim 1+(1+\beta+\vartheta)\int_{t-\tau}^t \mathbb{E}\|\sigma(X_s)\|_{HS}^2\mathrm{d}s\\
&\lesssim 1+\|\xi\|_\infty^2,\quad t\ge\tau.
\end{aligned}
$$

Carrying out the argument to deduce (2.69), we can derive that

$$
\mathbb{E}|X(t;\xi)-X(t;\eta)|^2\lesssim e^{-\lambda' t}\|\xi-\eta\|_\infty^2
$$

for some $\lambda'>0$. This, together with (2.64) and B–D–G's inequality, yields that

$$
\mathbb{E}\|X_t(\xi)-X_t(\eta)\|_\infty^2\lesssim e^{-\lambda' t}\|\xi-\eta\|_\infty^2.
$$

Finally, the desired assertion follows by imitating the corresponding argument of Theorem 2.18.

Remark 2.23 By a close scrutiny of the argument of Theorem 2.22, in fact, we can provide an upper bound of the Lipschitz constant $L>0$.

Chapter 3
Convergence Rate of Euler–Maruyama Scheme for FSDEs

In this chapter, we investigate convergence rate of the Euler–Maruyama (EM) scheme for a class of SDDEs, where the corresponding coefficients may be highly nonlinear w.r.t. the delay variables. More precisely, we reveal that convergence rate of the EM scheme is $\frac{1}{2}$ for the Brownian motion case. As for EM approximation for the pure jump case, it is best to consider the mean-square convergence. We show that the convergence rate is close to $\frac{1}{2}$. Moreover, under a logarithm Lipschitz condition we discuss convergence rate of the EM scheme associated with a range of FSDEs driven by jump processes. Sections 3.1–3.3 are mainly based on the results of [15] and Section 3.4 is based on the results in [13].

3.1 Introduction

Since most SDEs cannot be solved explicitly, numerical methods have become essential. Recently, there is an extensive literature on investigating strong convergence, weak convergence or sample path convergence of numerical schemes for SDEs; see, e.g., [61] for SDEs with a monotone condition, [67, 73, 112] for SDEs with jumps, [69, 73, 84, 100] for SDDEs and [67, 68] for SDEs with a one-sided Lipschitz condition. For the comprehensive treatment on numerical approximations of SDEs, we refer the readers to for example, [80, 112, 129]. Although there are numerous results on convergence of EM schemes, the study on convergence rate is relatively scarce. The recent work [62] reveals convergence rate of EM schemes for a class of SDEs under a Hölder condition, and, with local Lipschitz constants satisfying a logarithm growth condition, [154] and [9, 73] obtained convergence rate of EM approximations for SDEs and FSDEs with jumps, respectively.

In this chapter we study strong convergence rate of EM schemes for a class of SDDEs under a polynomial growth conditions, and for a range of FSDEs driven by jump processes under a logarithm Lipschitz condition, respectively. Aiming at deriving the convergence rate of the corresponding EM numerical schemes, the rest of this chapter is organized as follows: under highly nonlinear growth conditions w.r.t.

J. Bao et al., *Asymptotic Analysis for Functional Stochastic Differential Equations*, SpringerBriefs in Mathematics, DOI 10.1007/978-3-319-46979-9_3

the delay variables, in Section 3.2 we ascertain that convergence rate of EM schemes for SDDEs driven by Brownian motion is $\frac{1}{2}$, while in Section 3.3 we explore that it is best to consider the mean-square convergence of the associated EM approximation for the pure jump case, and show that the corresponding convergence rate is close to $\frac{1}{2}$. In the last section, under a logarithm Lipschitz condition we reveal that convergence rate of the EM scheme associated with a range of FSDEs driven by jump processes is close to $\frac{1}{2}$.

3.2 Convergence Rate under a Polynomial Condition: Systems Driven by Brownian Motions

In this section we are concerned with an SDDE on $\mathbb{R}^n$ in the form

$$dX(t) = b(X(t), X(t-\tau))dt + \sigma(X(t), X(t-\tau))dW(t), \ t > 0 \tag{3.1}$$

with the initial data $X(\theta) = \xi(\theta), \theta \in [-\tau, 0]$, where $b : \mathbb{R}^n \times \mathbb{R}^n \mapsto \mathbb{R}^n$, $\sigma : \mathbb{R}^n \times \mathbb{R}^n \mapsto \mathbb{R}^n \otimes \mathbb{R}^m$. Let $V : \mathbb{R}^n \times \mathbb{R}^n \mapsto \mathbb{R}_+$ such that

$$V(x, y) \leq K(1 + |x|^q + |y|^q), \quad x, y \in \mathbb{R}^n \tag{3.2}$$

for some $K > 0, q \geq 1$. For any $x_i, y_i \in \mathbb{R}^n$, $i = 1, 2$, we assume that there exists an $L > 0$ such that the following conditions hold:

(A1) $|b(x_1, y_1) - b(x_2, y_2)| \leq L|x_1 - x_2| + V(y_1, y_2)|y_1 - y_2|$;
(A2) $\|\sigma(x_1, y_1) - \sigma(x_2, y_2)\|_{HS} \leq L|x_1 - x_2| + V(y_1, y_2)|y_1 - y_2|$;
(A3) $|\xi(t) - \xi(s)| \leq L(s - t), \quad -\tau \leq t \leq s \leq 0$.

Remark 3.1 Clearly, if b and σ are globally Lipschitzian, then b and σ satisfy (A1) and (A2), respectively. On the other hand, we remark that b and σ may be *highly nonlinear* w.r.t. the delay variables. For example, for $\tilde{a}, \tilde{b}, \tilde{c} \in \mathbb{R}$, let $b(x, y) = \tilde{a}x + \tilde{b}y^3$, $\sigma(x, y) = \tilde{c}y^2$ for $x, y \in \mathbb{R}$. Then, (A1) and (A2) are verified by choosing $V(x, y) = 2(|\tilde{b}| \vee |\tilde{c}|)(1 + x^2 + y^2)$.

Fix a $T > 0$. We now introduce an EM algorithm for (3.1). Without loss of generality, we may assume that there exist sufficiently large integers $N, M > 0$ such that the stepsize $\delta := \tau/N = T/M \in (0, 1)$. The discrete-time EM scheme associated with (3.1) is defined by $\overline{Y}(t) = X(t) = \xi(t)$ for any $t \in [-\tau, 0]$, and, for any integer $k \geq 0$,

$$\begin{aligned}\overline{Y}((k+1)\delta) = \overline{Y}(k\delta) &+ b(\overline{Y}(k\delta), \overline{Y}(k\delta - \tau))\delta \\ &+ \sigma(\overline{Y}(k\delta), \overline{Y}(k\delta - \tau))\triangle W_{k\delta},\end{aligned} \tag{3.3}$$

where $\triangle W_{k\delta} := W((k+1)\delta) - W(k\delta)$, the increment of Brownian motion. For convenience of calculation, we define a continuous-time EM scheme corresponding to (3.1) by $Y(t) = X(t) = \xi(t)$ for $t \in [-\tau, 0]$, and

$$\mathrm{d}Y(t) = b(\overline{Y}(t_\delta), \overline{Y}(t_\delta - \tau))\mathrm{d}t + \sigma(\overline{Y}(t_\delta), \overline{Y}(t_\delta - \tau))\mathrm{d}W(t), \quad t > 0, \tag{3.4}$$

in which $t_\delta := \lfloor t/\delta \rfloor \delta$ with $\lfloor t/\delta \rfloor$ being the integer part of t/δ. If $Y(k\delta) = \overline{Y}(k\delta)$ for some integer $k \geq 0$, we deduce from (3.3) and (3.4) that

$$\begin{aligned} Y((k+1)\delta) &= Y(k\delta) + b(\overline{Y}(k\delta), \overline{Y}(k\delta - \tau))\delta + \sigma(\overline{Y}(k\delta), \overline{Y}(k\delta - \tau))\triangle W_{k\delta} \\ &= \overline{Y}(k\delta) + b(\overline{Y}(k\delta), \overline{Y}(k\delta - \tau))\delta + \sigma(\overline{Y}(k\delta), \overline{Y}(k\delta - \tau))\triangle W_{k\delta} \\ &= \overline{Y}((k+1)\delta). \end{aligned}$$

So, an induction argument leads to $Y(t_\delta) = \overline{Y}(t_\delta)$. That is, the discrete-time EM scheme coincides with the continuous-time EM scheme at the gridpoints.

Lemma 3.2 *Assume that* (A1)–(A3) *hold and let* $p \geq 2$. *Then,* (3.1) *admits a unique global strong solution* $(X(t))_{t \geq -\tau}$. *Moreover,*

$$\mathbb{E}\Big(\sup_{-\tau \leq t \leq T} |X(t)|^p\Big) \vee \mathbb{E}\Big(\sup_{-\tau \leq t \leq T} |Y(t)|^p\Big) \lesssim_T 1 + \|\xi\|_\infty^{p(1+q)}, \tag{3.5}$$

where $q \geq 1$ *was defined in* (3.2)*, and*

$$\mathbb{E}|Y(t) - \overline{Y}(t_\delta)|^p \lesssim_T \delta^{\frac{p}{2}}, \quad t \in [-\tau, T]. \tag{3.6}$$

Proof To begin, we establish existence and uniqueness of strong solutions to (3.1); see also Theorem A.4. For any $t \in [0, \tau]$, (3.1) reduces to an SDE without delay, and admits the following integral form

$$X(t) = \xi(0) + \int_0^t b(X(s), \xi(s - \tau))\mathrm{d}s + \int_0^t \sigma(X(s), \xi(s - \tau))\mathrm{d}W(s).$$

According to (A1) and (A2), for a frozen $y \in \mathbb{R}^n$, $b(\cdot, y)$ and $\sigma(\cdot, y)$ are global Lipschitz so that, by [99, Theorem 3.1, p. 51], (3.1) admits a unique strong solution $(X(t))_{t \in [0,\tau]}$. Next, we reconsider (3.1) on the time interval $[\tau, 2\tau]$ with the initial data $\eta(\theta) = X(\tau + \theta), \theta \in [-\tau, 0]$. Also, (3.1) becomes an SDE without delay, and admits the integral form below

$$X(t) = \eta(0) + \int_\tau^t b(X(s), \eta(s - 2\tau))\mathrm{d}s + \int_\tau^t \sigma(X(s), \eta(s - 2\tau))\mathrm{d}W(s).$$

Again, by virtue of [99, Theorem 3.1, p. 51], (3.1) possesses a unique strong solution $(X(t))_{t \in [0,2\tau]}$. Repeating the procedure above, we conclude that (3.1) enjoys a unique strong solution $(X(t))_{t \geq 0}$.

Next, we verify (3.5). Since the estimate of $\mathbb{E}(\sup_{-\tau\le t\le T}|X(t)|^p)$ is quite similar to that of $\mathbb{E}(\sup_{-\tau\le t\le T}|Y(t)|^p)$, in what follows, we only focus on the estimate of $\mathbb{E}(\sup_{-\tau\le t\le T}|Y(t)|^p)$ for the sake of brevity. By a straightforward computation, we deduce from (A1), (A2), and (3.2) that

$$|b(x,y)|+\|\sigma(x,y)\|_{HS}\lesssim 1+|x|+|y|^{1+q},\quad x,y\in\mathbb{R}^n. \tag{3.7}$$

To proceed, let $t\in[0,T]$ be arbitrary. By (3.7), in addition to Hölder's inequality and B–D–G's inequality, we derive that

$$\begin{aligned}
&\mathbb{E}\Big(\sup_{0\le s\le t}|Y(s)|^p\Big)\\
&\lesssim|\xi(0)|^p+\mathbb{E}\Big(\sup_{0\le s\le t}\Big|\int_0^s b(\overline{Y}(r_\delta),\overline{Y}(r_\delta-\tau))\mathrm{d}r\Big|^p\Big)\\
&\quad+\mathbb{E}\Big(\sup_{0\le s\le t}\Big|\int_0^s \sigma(\overline{Y}(r_\delta),\overline{Y}(r_\delta-\tau))\mathrm{d}W(r)\Big|^p\Big)\\
&\lesssim_T 1+\|\xi\|_\infty^p+\int_0^t\mathbb{E}\{|b(\overline{Y}(s_\delta),\overline{Y}(s_\delta-\tau))|^p+\|\sigma(\overline{Y}(s_\delta),\overline{Y}(s_\delta-\tau))\|_{HS}^p\}\mathrm{d}s\\
&\lesssim_T 1+\|\xi\|_\infty^p+\int_0^t\mathbb{E}|\overline{Y}(s_\delta)|^p\mathrm{d}s+\int_0^t\mathbb{E}|\overline{Y}(s_\delta-\tau)|^{p(1+q)}\mathrm{d}s\\
&\lesssim_T 1+\|\xi\|_\infty^p+\int_0^t\mathbb{E}|Y(s_\delta)|^p\mathrm{d}s+\int_0^t\mathbb{E}|Y(s_\delta-\tau)|^{p(1+q)}\mathrm{d}s,
\end{aligned}$$

where in the last step we have used $Y(t_\delta)=\bar{Y}(t_\delta)$. This, together with Gronwall's inequality, yields that

$$\mathbb{E}\Big(\sup_{0\le s\le t}|Y(s)|^p\Big)\lesssim_T 1+\|\xi\|_\infty^p+\int_0^t\mathbb{E}|Y(s_\delta-\tau)|^{p(1+q)}\mathrm{d}s. \tag{3.8}$$

Let

$$p_i=([T/\tau]+2-i)p(1+q)^{[T/\tau]+1-i},\quad i=1,2,\ldots,[T/\tau]+1.$$

Thus, due to $p\ge 2$, it is easy to see that $p_i\ge 2$ such that

$$p_{i+1}(1+q)<p_i\ \text{ and }\ p_{[T/\tau]+1}=p,\quad i=1,2,\ldots,[T/\tau]. \tag{3.9}$$

From (3.8), we obtain that

$$\mathbb{E}\Big(\sup_{-\tau\le s\le\tau}|Y(s)|^{p_1}\Big)\lesssim 1+\|\xi\|_\infty^{p_1(1+q)},$$

which, combining (3.8) with Hölder's inequality and (3.9), further implies

$$\begin{aligned}\mathbb{E}\Big(\sup_{-\tau\le s\le 2\tau}|Y(s)|^{p_2}\Big) &\lesssim_T 1+\|\xi\|_\infty^{p_2}+\int_0^{2\tau}\mathbb{E}|Y(s_\delta-\tau)|^{p_2(1+q)}\mathrm{d}s\\ &\lesssim_T 1+\|\xi\|_\infty^{p_2}+\int_{-\tau}^{\tau}(\mathbb{E}|Y(s_\delta)|^{p_1})^{\frac{p_2(1+q)}{p_1}}\mathrm{d}s\\ &\lesssim_T 1+\|\xi\|_\infty^{p_2(1+q)}.\end{aligned}$$

Repeating the previous procedure gives the desired assertion. Taking Hölder's inequality and [99, Theorem 7.1, p. 39] into account, we obtain from (3.7) that

$$\begin{aligned}\mathbb{E}|Y(t)-\overline{Y}(t_\delta)|^p &\lesssim \delta^{\frac{p-2}{2}}\int_{t_\delta}^t \mathbb{E}\{|b(Y(s_\delta),Y(s_\delta-\tau))|^p+\|\sigma(Y(s_\delta),Y(s_\delta-\tau))\|_{HS}^p\}\mathrm{d}s\\ &\lesssim \delta^{\frac{p}{2}}\int_{t_\delta}^t \mathbb{E}\{1+|Y(s_\delta)|^p+|Y(s_\delta-\tau)|^{p(1+q)}\}\mathrm{d}s,\quad t>0.\end{aligned}$$

Thus, (3.6) follows from (3.5) and (A3).

We now state our main result. Not only does it show strong convergence of the EM scheme associated with (3.1), but also reveals its convergence rate. Again, the drift and the diffusion coefficients may be highly nonlinear w.r.t. the delay variables.

Theorem 3.3 *Under* (A1)–(A3),

$$\mathbb{E}\Big(\sup_{0\le s\le T}|X(s)-Y(s)|^p\Big)\lesssim_T \delta^{p/2},\quad p>0,$$

that is, the convergence rate of the EM scheme (3.4) *is* 1/2.

Proof We adopt the so-called Yamada–Watanabe approximation method (see, e.g., [71]) to estimate the error. For fixed $\kappa>1$ and arbitrary $\varepsilon\in(0,1)$, there exists a continuous nonnegative function $\psi_{\kappa\varepsilon}(\cdot)$ with the support $[\varepsilon/\kappa,\varepsilon]$ such that

$$\int_{\varepsilon/\kappa}^{\varepsilon}\psi_{\kappa\varepsilon}(x)\mathrm{d}x=1 \;\; and \;\; \psi_{\kappa\varepsilon}(x)\le\frac{2}{x\ln\kappa},\quad x>0.$$

Define

$$\phi_{\kappa\varepsilon}(x)=\int_0^x\int_0^y\psi_{\kappa\varepsilon}(z)\mathrm{d}z\mathrm{d}y,\quad x>0.$$

Then, $\phi_{\kappa\varepsilon}\in C^2(\mathbb{R}_+;\mathbb{R}_+)$ possesses the following properties:

$$x-\varepsilon\le\phi_{\kappa\varepsilon}(x)\le x,\quad x>0, \tag{3.10}$$

and

$$0 \le \phi'_{\kappa\varepsilon}(x) \le 1, \quad \phi''_{\kappa\varepsilon}(x) \le \frac{2}{x \ln \kappa} \mathbf{1}_{[\varepsilon/\kappa, \varepsilon]}(x), \quad x > 0. \tag{3.11}$$

Set

$$V_{\kappa\varepsilon}(x) := \phi_{\kappa\varepsilon}(|x|), \quad x \in \mathbb{R}^n. \tag{3.12}$$

Then, one has $V_{\kappa\varepsilon} \in C^2(\mathbb{R}^n; \mathbb{R}_+)$,

$$(\nabla V_{\kappa\varepsilon})(x) = \phi'_{\kappa\varepsilon}(|x|)|x|^{-1}x,$$

and

$$(\nabla^2 V_{\kappa\varepsilon})(x) = \phi'_{\kappa\varepsilon}(|x|)(|x|^2 I_{n\times n} - x \otimes x)|x|^{-3} + |x|^{-2}\phi''_{\kappa\varepsilon}(|x|)x \otimes x,$$

where ∇ and ∇^2 denote the gradient and Hessian operators, respectively, and $x \otimes x = xx^*$ with x^* being the transpose of $x \in \mathbb{R}^n$. Moreover, we find that

$$0 \le |(\nabla V_{\kappa\varepsilon})(x)| \le 1 \text{ and } \|(\nabla^2 V_{\kappa\varepsilon})(x)\| \le 2n\Big(1 + \frac{1}{\ln \kappa}\Big)\frac{1}{|x|}\mathbf{1}_{[\varepsilon/\kappa, \varepsilon]}(|x|). \tag{3.13}$$

For any $t \in [-\tau, T]$, let

$$Z(t) = X(t) - Y(t),\ \overline{Z}(t) = Y(t) - \overline{Y}(t_\delta) \text{ and } \widetilde{Z}(t) = (X(t), \overline{Y}(t_\delta)) \in \mathbb{R}^{2n}. \tag{3.14}$$

An application of the Itô formula yields that

$$\begin{aligned} V_{\kappa\varepsilon}(Z(t)) &= \int_0^t \langle (\nabla V_{\kappa\varepsilon})(Z(s)), \Gamma_1(s)\rangle \mathrm{d}s + \int_0^t \langle (\nabla V_{\kappa\varepsilon})(Z(s)), \Gamma_2(s)\mathrm{d}W(s)\rangle \\ &\quad + \frac{1}{2}\int_0^t \mathrm{trace}\{(\Gamma_2(s))^*(\nabla^2 V_{\kappa\varepsilon})(Z(s))\Gamma_2(s)\}\mathrm{d}s \\ &=: I_1(t) + I_2(t) + I_3(t), \end{aligned}$$

where, for any $t \in [0, T]$,

$$\Gamma_1(t) := b(X(t), X(t-\tau)) - b(\overline{Y}(t_\delta), \overline{Y}(t_\delta - \tau)), \tag{3.15}$$

and $\Gamma_2(t) := \sigma(X(t), X(t-\tau)) - \sigma(\overline{Y}(t_\delta), \overline{Y}(t_\delta - \tau))$. For any $p \ge 1$, observe from (3.2) and (3.5) that

$$\begin{aligned} \sup_{-\tau\le t\le T} \mathbb{E}V(\widetilde{Z}(t))^p &\lesssim 1 + \sup_{-\tau\le t\le T} \mathbb{E}|X(t)|^{pq} + \sup_{-\tau\le t\le T} \mathbb{E}|\overline{Y}(t_\delta)|^{pq} \\ &\lesssim_T 1 + \|\xi\|_\infty^{pq(1+q)}. \end{aligned} \tag{3.16}$$

By (3.13), (A1)–(A2), Hölder's inequality and B–D–G's inequality, we derive from (3.6) and (3.16) that

$$
\begin{aligned}
&\mathbb{E}\Big(\sup_{0\le s\le t}|I_1(s)|^p\Big)+\mathbb{E}\Big(\sup_{0\le s\le t}|I_3(s)|^p\Big)\\
&\lesssim_T \int_0^t \mathbb{E}|\Gamma_1(s)|^p\mathrm{d}s+\mathbb{E}\Big(\int_0^t \|\Gamma_2(s)\|_{HS}^2\mathrm{d}s\Big)^{\frac{p}{2}}\\
&\lesssim_T \int_0^t \{\mathbb{E}|\Gamma_1(s)|^p+\mathbb{E}|\Gamma_2(s)|^p\}\mathrm{d}s \qquad (3.17)\\
&\lesssim_T \int_0^t \{\mathbb{E}|Z(s)|^p+(\mathbb{E}V(\widetilde{Z}(s-\tau))^{2p})^{\frac{1}{2}}(\mathbb{E}|Z(s-\tau)|^{2p})^{\frac{1}{2}}\\
&\quad+\mathbb{E}|\overline{Z}(s)|^p+(\mathbb{E}V(\widetilde{Z}(s-\tau))^{2p})^{\frac{1}{2}}(\mathbb{E}|\overline{Z}(s-\tau)|^{2p})^{\frac{1}{2}}\}\mathrm{d}s\\
&\lesssim_T \int_0^t \{\delta^{p/2}+\mathbb{E}|Z(s)|^p+(\mathbb{E}|Z(s-\tau)|^{2p})^{\frac{1}{2}}\}\mathrm{d}s,
\end{aligned}
$$

and that

$$
\begin{aligned}
\mathbb{E}\Big(\sup_{0\le s\le t}|I_2(s)|^p\Big)&\lesssim_T \int_0^t \mathbb{E}\{\|(\nabla^2 V_{\kappa\varepsilon})(Z(s))\|\cdot\|\Gamma_2(s)\|_{HS}^2\}^p\mathrm{d}s\\
&\lesssim_T \mathbb{E}\int_0^t \frac{1}{|Z(s)|^p}\{|Z(s)|^{2p}+V(\widetilde{Z}(s-\tau))^{2p}|Z(s-\tau)|^{2p}\\
&\quad+|\overline{Z}(s)|^{2p}+V(\widetilde{Z}(s-\tau))^{2p}|\overline{Z}(s-\tau)|^{2p}\}\mathbf{1}_{[\varepsilon/\delta,\varepsilon]}(|Z(s)|)\mathrm{d}s\\
&\lesssim_T \int_0^t \{\mathbb{E}|Z(s)|^p+\varepsilon^{-p}(\mathbb{E}V(\widetilde{Z}(s-\tau))^{4p})^{\frac{1}{2}}(\mathbb{E}|Z(s-\tau)|^{4p})^{\frac{1}{2}}\\
&\quad+\varepsilon^{-p}\mathbb{E}|\overline{Z}(s)|^{2p}+\varepsilon^{-p}(\mathbb{E}V(\widetilde{Z}(s-\tau))^{4p})^{\frac{1}{2}}(\mathbb{E}|\overline{Z}(s-\tau)|^{4p})^{\frac{1}{2}}\}\mathrm{d}s\\
&\lesssim_T \int_0^t \{\varepsilon^{-p}\delta^p+\mathbb{E}|Z(s)|^p+\varepsilon^{-p}(\mathbb{E}|Z(s-\tau)|^{4p})^{\frac{1}{2}}\}\mathrm{d}s.
\end{aligned}
$$

Then, besides (3.10), we arrive at

$$
\begin{aligned}
\mathbb{E}\Big(\sup_{0\le s\le t}|Z(s)|^p\Big)&\lesssim \varepsilon^p+\mathbb{E}\Big(\sup_{0\le s\le t}V_{\kappa\varepsilon}(Z(s))^p\Big)\\
&\lesssim_T \varepsilon^p+\delta^{p/2}+\varepsilon^{-p}\delta^p+\int_0^t \mathbb{E}|Z(s)|^p\mathrm{d}s\\
&\quad+\int_0^{(t-\tau)\vee 0}(\mathbb{E}|Z(s)|^{2p})^{\frac{1}{2}}\mathrm{d}s+\varepsilon^{-p}\int_0^{(t-\tau)\vee 0}(\mathbb{E}|Z(s)|^{4p})^{\frac{1}{2}}\mathrm{d}s,
\end{aligned}
$$

where in the last step, we used $Z(\theta) = 0$ for $\theta \in [-\tau, 0]$. This, together with Gronwall's inequality, implies

$$\begin{aligned}\mathbb{E}\Big(\sup_{0\le s\le t}|Z(s)|^p\Big) \lesssim_T \varepsilon^p + \delta^{p/2} + \varepsilon^{-p}\delta^p + \int_0^{(t-\tau)\vee 0}(\mathbb{E}|Z(s)|^{2p})^{\frac{1}{2}}\mathrm{d}s\\ + \varepsilon^{-p}\int_0^{(t-\tau)\vee 0}(\mathbb{E}|Z(s)|^{4p})^{\frac{1}{2}}\mathrm{d}s.\end{aligned} \tag{3.18}$$

For any $p \ge 2$, let

$$p_i = p([T/\tau] + 2 - i)4^{[T/\tau]+1-i}, \quad i = 1, 2, \ldots, [T/\tau] + 1.$$

It is easy to see that

$$4p_{i+1} < p_i \text{ and } p_{[T/\tau]+1} = p, \quad i = 1, 2, \ldots, [T/\tau]. \tag{3.19}$$

Taking $\varepsilon = \delta^{\frac{1}{2}}$ in (3.18), we obtain that

$$\mathbb{E}\Big(\sup_{0\le s\le \tau}|Z(s)|^{p_1}\Big) \lesssim \delta^{\frac{p_1}{2}}.$$

This, combined with (3.19) and Hölder's inequality, further gives that

$$\begin{aligned}\mathbb{E}\Big(\sup_{0\le s\le 2\tau}|Z(s)|^{p_2}\Big) &\lesssim_T \delta^{\frac{p_2}{2}} + \int_0^{\tau}(\mathbb{E}|Z(s)|^{p_1})^{\frac{p_2}{p_1}}\mathrm{d}s + \delta^{-\frac{p_2}{2}}\int_0^{\tau}(\mathbb{E}|Z(s)|^{p_1})^{\frac{2p_2}{p_1}}\mathrm{d}s\\ &\lesssim_T \delta^{\frac{p_2}{2}}\end{aligned}$$

by taking $\varepsilon = \delta^{\frac{1}{2}}$ in (3.18). The desired assertion then follows by repeating the previous procedure.

3.3 Convergence Rate: Polynomial Condition for Systems Driven by Pure Jump Processes

In the previous section, we obtained strong convergence of the EM scheme for a class of SDDEs, and revealed the corresponding convergence rate is $1/2$ although both the drift and the diffusion coefficients may be highly nonlinear w.r.t. the delay variables. In this section we turn to the counterpart for SDDEs with pure jumps.

In this section we consider an SDDE with jumps in the form

$$\mathrm{d}X(t) = b(X(t), X(t-\tau))\mathrm{d}t + \int_{\Gamma} h(X(t-), X((t-\tau)-), u)\widetilde{N}(\mathrm{d}t, \mathrm{d}u) \tag{3.20}$$

with the initial data $X(\theta) = \xi(\theta), \theta \in [-\tau, 0]$, where $b : \mathbb{R}^n \times \mathbb{R}^n \mapsto \mathbb{R}^n, h : \mathbb{R}^n \times \mathbb{R}^n \times \Gamma \mapsto \mathbb{R}^n$. Besides (A1) and (A3), we assume that there exists an $L' > 0$ such that

(A4) $|h(x_1, y_1, u) - h(x_2, y_2, u)| \le L'(|x_1 - x_2| + V(y_1, y_2)|y_1 - y_2|)|u|$ for any $x_i, y_i \in \mathbb{R}^n, i = 1, 2$ and $u \in \Gamma$, in which $V(\cdot, \cdot)$ such that (3.2) holds.

Remark 3.4 The jump coefficient may be *highly nonlinear* w.r.t. the delay argument. For instance, for $x, y \in \mathbb{R}, u \in \Gamma$, and $q > 1$, $h(x, y, u) = y^q u$ satisfies (A4).

The discrete-time EM scheme associated with (3.20) is defined by $\overline{Y}(t) = X(t) = \xi(t)$ for $t \in [-\tau, 0]$, and, for any integer $k \ge 0$,

$$\begin{aligned}\overline{Y}((k+1)\delta) = \overline{Y}(k\delta) + b(\overline{Y}(k\delta), \overline{Y}(k\delta - \tau))\delta \\ + \int_\Gamma h(\overline{Y}(k\delta), \overline{Y}(k\delta - \tau), u)\Delta_{k\delta}\widetilde{N}(du),\end{aligned} \tag{3.21}$$

where $\Delta_{k\delta}\widetilde{N}(du) := \widetilde{N}((0, (k+1)\delta], du) - \widetilde{N}((0, k\delta], du)$. For convenience, we also define a continuous-time EM scheme for (3.20) as $Y(t) = X(t) = \xi(t), t \in [-\tau, 0]$, and

$$dY(t) = b(\overline{Y}(t_\delta), \overline{Y}(t_\delta - \tau))dt + \int_\Gamma h(\overline{Y}(t_\delta), \overline{Y}(t_\delta - \tau), u)\widetilde{N}(dt, du), \quad t > 0. \tag{3.22}$$

By a straightforward calculation, one has $Y(t_\delta) = \overline{Y}(t_\delta)$. To reveal the convergence order of EM scheme (3.22), we need two auxiliary lemmas, where the first one is concerned with Bichteler–Jacod inequality for Poisson integrals (see, e.g., [3, p. 265], [101, Lemma 3.1] and [103, Theorem 1]).

Lemma 3.5 *Let $\Phi : \mathbb{R}_+ \times \Gamma \times \Omega \mapsto \mathbb{R}^n$ and assume that*

$$\int_0^t \int_\Gamma \mathbb{E}|\Phi(s, u)|^p \nu(du)ds < \infty, \quad t \ge 0, \quad p \ge 2.$$

Then, there exists a $c_p > 0$ such that

$$\begin{aligned}\mathbb{E}\Big(\sup_{0\le s\le t}\Big|\int_0^s \int_\Gamma \Phi(r, u)\tilde{N}(du, dr)\Big|^p\Big) \le c_p\Big\{\mathbb{E}\Big(\int_0^t \int_\Gamma |\Phi(s, u)|^2 \nu(du)ds\Big)^{\frac{p}{2}} \\ + \int_0^t \int_\Gamma \mathbb{E}|\Phi(s, u)|^p \nu(du)ds\Big\}.\end{aligned}$$

Lemma 3.6 *Let* (A1), (A3), *and* (A4) *hold and further assume* $\int_U |u|^p \nu(\mathrm{d}u) < \infty$ *for any* $p \geq 2$. *Then,* (3.20) *has a unique global solution* $(X(t))_{t\geq -\tau}$. *Moreover, for any* $p \geq 2$

$$\mathbb{E}\Big(\sup_{-\tau\leq t\leq T}|X(t)|^p\Big) \vee \mathbb{E}\Big(\sup_{-\tau\leq t\leq T}|Y(t)|^p\Big) \lesssim_T 1 + \|\xi\|_\infty^{p(1+q)}, \tag{3.23}$$

and

$$\mathbb{E}|Y(t) - \overline{Y}(t_\delta)|^p \lesssim_T \delta, \quad t \in [-\tau, T]. \tag{3.24}$$

Proof Since the argument of Lemma 3.6 is quite similar to that of Lemma 3.2, we only give an outline of the proof to point out some differences. Following the argument of Lemma 3.2 on existence and uniqueness of solutions and taking [130, Theorem 117, p. 79] (see also Theorem A.8) into consideration, we conclude that (3.20) admits a unique global solution $(X(t))_{t\geq -\tau}$. As we explain before, hereinafter we only estimate

$$\mathbb{E}\Big(\sup_{-\tau\leq t\leq T}|Y(t)|^p\Big) \lesssim_T 1 + \|\xi\|_\infty^{p(1+q)}. \tag{3.25}$$

Using Hölder's inequality, and Lemma 3.5, we deduce from (A1) and (A4) that

$$\begin{aligned}
\mathbb{E}\Big(\sup_{0\leq s\leq t}|Y(s)|^p\Big) &\lesssim_T |\xi(0)|^p + \int_0^t \mathbb{E}|b(\overline{Y}(r_\delta), \overline{Y}(r_\delta - \tau))|^p \mathrm{d}r \\
&\quad + \mathbb{E}\Big(\int_0^t \int_\Gamma |h(\overline{Y}(r_\delta), \overline{Y}(r_\delta - \tau), u)|^2 \nu(\mathrm{d}u)\mathrm{d}r\Big)^{p/2} \\
&\quad + \int_0^t \int_\Gamma \mathbb{E}|h(\overline{Y}(r_\delta), \overline{Y}(r_\delta - \tau), u)|^p \nu(\mathrm{d}u)\mathrm{d}r \\
&\lesssim_T 1 + \|\xi\|_\infty^p + \int_0^t \{\mathbb{E}|Y(s_\delta)|^p + \mathbb{E}|Y(s_\delta - \tau)|^{p(1+q)}\}\mathrm{d}s.
\end{aligned}$$

Thus, Gronwall's inequality and an induction argument give (3.25). In view of Hölder's inequality, and Lemma 3.5, together with (A1) and (A4), we observe that

$$\begin{aligned}
\mathbb{E}|Y(t) - \overline{Y}(t_\delta)|^p &\lesssim \delta^{p-1} \int_{t_\delta}^t \mathbb{E}|b(\overline{Y}(s_\delta), \overline{Y}(s_\delta - \tau))|^p \mathrm{d}s \\
&\quad + \int_{t_\delta}^t \int_\Gamma \mathbb{E}|h(\overline{Y}(s_\delta), \overline{Y}(s_\delta - \tau), u)|^p \nu(\mathrm{d}u)\mathrm{d}s \\
&\quad + \mathbb{E}\Big(\int_{t_\delta}^t \int_\Gamma |h(\overline{Y}(s_\delta), \overline{Y}(s_\delta - \tau), u)|^2 \nu(\mathrm{d}u)\mathrm{d}s\Big)^{p/2} \\
&\lesssim \int_{t_\delta}^t \{\mathbb{E}|Y(s_\delta)|^p + \mathbb{E}|Y(s_\delta - \tau)|^{p(1+q)}\}\mathrm{d}s.
\end{aligned}$$

Then, (3.24) follows from (3.23) and (A3).

Remark 3.7 We remark that, for $p \geq 2$, all p-th moments of $Y(t) - \overline{Y}(t_\delta)$ are bounded by δ up to a constant, which is completely different from the Brownian motion case (i.e., (3.6)). This is due to the fact that all moments of the increment $\widetilde{N}((0, (k+1)\delta, du) - \widetilde{N}((0, k\delta], du)$ have order $O(\delta)$ for $\delta \in (0, 1)$.

We now state our main result in this section.

Theorem 3.8 *Under the assumptions of Lemma 3.6, for any $p \geq 2$ and $\theta, \alpha \in (0, 1)$,*

$$\mathbb{E}\Big(\sup_{0 \leq s \leq T} |X(s) - Y(s)|^p\Big) \lesssim_T \delta^{\frac{1}{(1+\theta)^{\lceil T/\tau \rceil}(1+\alpha)}}.$$

Proof The proof of Theorem 3.8 is similar to that of Theorem 3.3. We give a sketch of the proof to highlight the corresponding differences. Let $Z(t), \bar{Z}(t), \tilde{Z}(t)$ be defined as in (3.14), and $V_{\kappa\varepsilon} \in C^2(\mathbb{R}^n; \mathbb{R}_+)$ be defined as in (3.12). The Itô formula and the Taylor expansion give that for any $t \subset [0, T]$

$$\begin{aligned}
&V_{\kappa\varepsilon}(Z(t)) \\
&= V_{\kappa\varepsilon}(Z(0)) + \int_0^t \langle (\nabla V_{\kappa\varepsilon})(Z(s)), \Gamma_1(s) \rangle \mathrm{d}s \\
&+ \int_0^t \int_\Gamma \langle (\nabla V_{\kappa\varepsilon})(Z(s) + \xi(s)\Gamma_0(s, u)) - (\nabla V_{\kappa\varepsilon})(Z(s)), \Gamma_0(s, u) \rangle \nu(\mathrm{d}u)\mathrm{d}s \\
&+ \int_0^t \int_\Gamma \langle (\nabla V_{\kappa\varepsilon})(Z(s) + \xi(s)\Gamma_0(s, u)), \Gamma_0(s, u) \rangle \widetilde{N}(\mathrm{d}u, \mathrm{d}s)
\end{aligned}$$

for some random variable $\xi \in [0, 1]$, where $\Gamma_1(\cdot)$ is defined as in (3.15) and

$$\Gamma_0(t, u) := h(X(t), X(t-\tau), u) - h(\overline{Y}(t_\delta), \overline{Y}(t_\delta - \tau), u), \quad t \in [0, T], u \in \Gamma.$$

This, together with (3.10) and (3.13), yields that

$$\begin{aligned}
|Z(t)| &\leq \varepsilon + V_{\delta\varepsilon}(Z(t)) \\
&\leq \varepsilon + \int_0^t |\Gamma_1(s)| \mathrm{d}s + 2\int_0^t \int_\Gamma |\Gamma_0(s, u)| \nu(\mathrm{d}u)\mathrm{d}s \\
&\quad + \int_0^t \int_\Gamma \langle (\nabla V_{\kappa\varepsilon})(Z(s) + \xi(s)\Gamma_0(s, u)), \Gamma_0(s, u) \rangle \widetilde{N}(\mathrm{d}u, \mathrm{d}s).
\end{aligned}$$

For any $p > 0$, note from (3.2) and (3.23) that

$$\sup_{-\tau \leq t \leq T} \mathbb{E} V(\widetilde{Z}(t))^p \lesssim 1 + \|\xi\|_\infty^{pq(1+q)}.$$

Consequently, for any $p \geq 2$ and $t \in [0, T]$, using (3.13), (3.24), Lemma 3.5, and the Hölder inequality, we derive from (A1) and (A4) that

$$
\begin{aligned}
\mathbb{E}\Big(\sup_{0\le s\le t}|Z(s)|^p\Big) &\lesssim \varepsilon^p + \mathbb{E}\Big(\sup_{0\le s\le t} V_{\delta\varepsilon}(Z(s))^p\Big) \\
&\lesssim \varepsilon^p + \int_0^t \mathbb{E}|\Gamma_1(s)|^p \mathrm{d}s + \int_0^t\int_\Gamma \mathbb{E}|\Gamma_0(s,u)|^p \nu(\mathrm{d}u)\mathrm{d}s \\
&\quad + \mathbb{E}\Big(\int_0^t\int_\Gamma |\Gamma_0(s,u)|^2\nu(\mathrm{d}u)\mathrm{d}s\Big)^{\frac{p}{2}} \\
&\lesssim \varepsilon^p + \int_0^t \mathbb{E}[|X(s)-\overline{Y}(s)| + V(\widetilde{Z}(s-\tau))|X(s-\tau)-\overline{Y}(s-\tau)|]^p \mathrm{d}s \\
&\lesssim \varepsilon^p + \int_0^t \{\mathbb{E}|Z(s)|^p + \mathbb{E}|\overline{Z}(s)|^p + \mathbb{E}(V(\widetilde{Z}(s-\tau))^p|Z(s-\tau)|^p) \\
&\quad + \mathbb{E}(V(\widetilde{Z}(s-\tau))^p|\overline{Z}(s-\tau)|^p)\}\mathrm{d}s \\
&\lesssim \varepsilon^p + \int_0^t \mathbb{E}|Z(s)|^p\mathrm{d}s + \int_0^{(t-\tau)\vee 0} (\mathbb{E}|Z(s)|^{p(1+\theta)})^{\frac{1}{1+\theta}}\mathrm{d}s \\
&\quad + \int_0^t \mathbb{E}|\overline{Z}(s)|^p\mathrm{d}s + \int_{-\tau}^{t-\tau} (\mathbb{E}|\overline{Z}(s)|^{p(1+\theta)})^{\frac{1}{1+\theta}}\mathrm{d}s \\
&\lesssim \varepsilon^p + \delta^{\frac{1}{1+\theta}} + \int_0^t \mathbb{E}|Z(s)|^p\mathrm{d}s + \int_0^{(t-\tau)\vee 0} (\mathbb{E}|Z(s)|^{p(1+\theta)})^{\frac{1}{1+\theta}}\mathrm{d}s,
\end{aligned}
$$

where $\theta \in (0, 1)$ is an arbitrary constant. An application of the Gronwall inequality then gives that

$$
\mathbb{E}\Big(\sup_{0\le s\le t}|Z(s)|^p\Big) \lesssim \delta + \delta^{\frac{1}{1+\theta}} + \int_0^{(t-\tau)\vee 0} (\mathbb{E}|Z(s)|^{p(1+\theta)})^{\frac{1}{1+\theta}}\mathrm{d}s \tag{3.26}
$$

by taking $\varepsilon = \delta^{\frac{1}{p}}$. For $\theta \in (0, 1)$ in (3.26) and any $\alpha \in (0, 1)$, let

$$
p_i = p(1+\theta)^{([T/\tau]+1-i)(1+\alpha)}, \quad i = 1, 2, \ldots, [T/\tau]+1.
$$

It is easily seen that

$$
(1+\theta)p_{i+1} < p_i \text{ and } p_{[T/\tau]+1} = p, \quad i = 1, 2, \ldots, [T/\tau]. \tag{3.27}
$$

By (3.26), we get

$$
\mathbb{E}\Big(\sup_{0\le s\le \tau}|Z(s)|^{p_1}\Big) \lesssim \delta.
$$

This, together with (3.26) and the Hölder inequality, yields that

$$\begin{aligned}\mathbb{E}\Big(\sup_{0\le s\le 2\tau}|Z(s)|^{p_2}\Big)&\lesssim \delta+\delta^{\frac{1}{1+\theta}}+\int_0^\tau(\mathbb{E}|Z(s)|^{p_2(1+\theta)})^{\frac{1}{1+\theta}}\mathrm{d}s\\&\lesssim \delta+\delta^{\frac{1}{1+\theta}}+\int_0^\tau(\mathbb{E}|Z(s)|^{p_1})^{\frac{p_2}{p_1}}\mathrm{d}s\\&\lesssim \delta+\delta^{\frac{1}{1+\theta}}+\delta^{\frac{p_2}{p_1}}\\&\lesssim \delta^{\frac{p_2}{p_1}},\end{aligned}\tag{3.28}$$

where the last step is due to (3.27). Similarly, we have from (3.26)–(3.28) that

$$\begin{aligned}\mathbb{E}\Big(\sup_{0\le s\le 3\tau}|Z(s)|^{p_3}\Big)&\lesssim \delta+\delta^{\frac{1}{1+\theta}}+\int_0^{2\tau}(\mathbb{E}|Z(s)|^{p_3(1+\theta)})^{\frac{1}{1+\theta}}\mathrm{d}s\\&\lesssim \delta+\delta^{\frac{1}{1+\theta}}+\int_0^{2\tau}(\mathbb{E}|Z(s)|^{p_2})^{\frac{p_3}{p_2}}\mathrm{d}s\\&\lesssim \delta+\delta^{\frac{1}{1+\theta}}+\delta^{\frac{p_3}{p_1}}\\&\lesssim \delta^{\frac{p_3}{p_1}}.\end{aligned}$$

As a result, the desired assertion follows from an induction argument.

Remark 3.9 By Theorem 3.8, with $p\ge 2$ increasing the convergence rate of EM scheme (3.22) is decreasing, which is quite different from the Brownian motion case with a constant order $\frac{1}{2}$, and it is therefore best to use the mean-square convergence for the jump case. On the other hand, we point out that the order of mean-square convergence is close to $\frac{1}{2}$ although the jump diffusion may be highly nonlinear w.r.t. the delay variables.

Remark 3.10 By a close inspection of the proof for Theorem 3.8, the conditions (A4) and $\int_U|u|^p\nu(\mathrm{d}u)<\infty$ for any $p\ge 2$ can be replaced by: for any $p>2$ there exist $K_p, K_0>0$ and $q_1,q_2>1$ such that for any $x,y,\bar{x},\bar{y}\in\mathbb{R}^n$,

$$\begin{aligned}&\Big(\int_\Gamma|h(x,y,u)-h(\bar{x},\bar{y},u)|^2\nu(\mathrm{d}u)\Big)^{p/2}+\int_\Gamma|h(x,y,u)-h(\bar{x},\bar{y},u)|^p\nu(\mathrm{d}u)\\&\le K_p(|x-\bar{x}|^p+(1+|y|^{q_1}+|\bar{y}|^{q_2})|y-\bar{y}|^p),\end{aligned}$$

and

$$\Big(\int_\Gamma|h(x,y,u)|^2\nu(\mathrm{d}u)\Big)^{p/2}+\int_\Gamma|h(x,y,u)|^p\nu(\mathrm{d}u)\le K_0(1+|x|^p+|y|^{q_2}).$$

3.4 Convergence Rate under a Logarithm Lipschitz Condition

In this section, we consider the following FSDE with jumps

$$\mathrm{d}X(t) = b(X_t)\mathrm{d}t + \sigma(X_t)\mathrm{d}W(t) + \int_\Gamma \gamma(X_{t-}, u)\widetilde{N}(\mathrm{d}t, \mathrm{d}u), \quad t > 0 \tag{3.29}$$

with the initial data $X_0 = \xi$, where $b : \mathscr{D} \mapsto \mathbb{R}^n$, $\sigma : \mathscr{D} \mapsto \mathbb{R}^n \otimes \mathbb{R}^m$, and $\gamma : \mathscr{D} \times \Gamma \mapsto \mathbb{R}^n$. Throughout this section, we suppose that $W(t)$ and $N(\mathrm{d}t, \mathrm{d}u)$ are independent. Let $\mu(\cdot)$ be a probability measure on $[-\tau, 0]$. In addition to (A3), we further assume that for any $\xi, \eta \in \mathscr{D}$ and $p \geq 2$, the following conditions hold.

(H1) There exists an $L > 0$ such that

$$|b(\xi) - b(\eta)|^2 \vee \|\sigma(\xi) - \sigma(\eta)\|_{HS}^2 \leq L \int_{[-\tau,0]} |\xi(\theta) - \eta(\theta)|^2 \mu(\mathrm{d}\theta).$$

(H2) There exists a $K^{(p)} > 0$ such that

$$\left(\int_U |\gamma(\xi, u) - \gamma(\eta, u)|^2 \nu(\mathrm{d}u)\right)^{p/2} + \int_U |\gamma(\xi, u) - \gamma(\eta, u)|^p \nu(\mathrm{d}u)$$
$$\leq K^{(p)} \int_{[-\tau,0]} |\xi(\theta) - \eta(\theta)|^p \mu(\mathrm{d}\theta).$$

(H3)

$$\left(\int_U |\gamma(0, u)|^2 \nu(du)\right)^{p/2} + \int_U |\gamma(0, u)|^p \nu(\mathrm{d}u) < \infty.$$

For a given $T > 0$, the time-step size $\delta \in (0, 1)$ is defined by $\delta = \frac{\tau}{N} = \frac{T}{M}$ for some integers $N > \tau$ and $M > T$. The EM method applied to (3.29) produces approximations $\overline{Y}(k\delta) \approx X(k\delta)$ by setting $\overline{Y}(k\delta) = X(k\delta) = \xi(k\delta)$, $-N \leq k \leq 0$, and for any $k \geq 0$,

$$\overline{Y}((k+1)\delta) = \overline{Y}(k\delta) + b(\overline{Y}_{k\delta})\delta + \sigma(\overline{Y}_{k\delta})\triangle W_{k\delta} + \int_\Gamma \gamma(\overline{Y}_{k\delta}, u)\triangle_{k\delta}\widetilde{N}(\mathrm{d}u), \tag{3.30}$$

where $\triangle W_{k\delta}$ and $\triangle_{k\delta}\widetilde{N}(\mathrm{d}u)$ are defined as in (3.3) and (3.21), respectively, and $\overline{Y}_{k\delta} = \{\overline{Y}_{k\delta}(\theta) : -\tau \leq \theta \leq 0\}$, a $\mathscr{D}$-valued random variable, is defined by

$$\overline{Y}_{k\delta}(\theta) = \frac{(i+1)\delta - \theta}{\delta}\overline{Y}((k+i)\delta) + \frac{\theta - i\delta}{\delta}\overline{Y}((k+i+1)\delta) \tag{3.31}$$

for $i\delta \leq \theta \leq (i+1)\delta$, $i = -N, \ldots, -1$. Given the discrete-time approximation $(\overline{Y}(k\delta))_{k\geq 0}$, we define a continuous-time approximation $(Y(t))_{t\geq 0}$ by $Y(t) = X(t) = \xi(t)$, $-\tau \leq t \leq 0$, while for $t > 0$, by

$$Y(t) = \xi(0) + \int_0^t b(\overline{Y}_{s_\delta})\mathrm{d}s + \int_0^t \sigma(\overline{Y}_{s_\delta})\mathrm{d}W(s) + \int_0^t \int_\Gamma \gamma(\overline{Y}_{s_\delta}, u)\widetilde{N}(\mathrm{d}s, \mathrm{d}u). \tag{3.32}$$

It is readily seen that $Y(t_\delta) = \overline{Y}(t_\delta)$ for arbitrary $t \geq 0$.

Lemma 3.11 *Under* (H1) *and* (H2),

$$\mathbb{E}\Big(\sup_{-\tau\leq t\leq T} |X(t)|^p\Big) \vee \mathbb{E}\Big(\sup_{-\tau\leq t\leq T} |Y(t)|^p\Big) \lesssim_T 1 + \|\xi\|_\infty^p.$$

Proof Since the arguments of the uniform p-th moment bounds for the exact and continuous approximate solutions to (3.29) are very similar, we need only give an estimate for the continuous-time approximate solution $Y(t)$. Due to $Y(t_\delta) = \overline{Y}(t_\delta)$, it follows from (3.31) that

$$\|\overline{Y}_{t_\delta}\|_\infty \leq \sup_{-\tau\leq s\leq t} |Y(s)|. \tag{3.33}$$

Next, for any $p \geq 2$ and $\xi \in \mathscr{D}$, from (H1) and (H2), one has

$$\begin{aligned} &|b(\xi)|^2 + \|\sigma(\xi)\|_{HS}^2 \lesssim 1 + \|\xi\|_\infty^2, \quad \text{and} \\ &\Big(\int_\Gamma |\gamma(\xi, u)|^2 \nu(\mathrm{d}u)\Big)^{p/2} + \int_\Gamma |\gamma(\xi, u)|^p \nu(\mathrm{d}u) \lesssim_p 1 + \|\xi\|_\infty^p. \end{aligned} \tag{3.34}$$

Thus, for any $t \in [0, T]$, applying Hölder's inequality, B–D–G's inequality, and Lemma 3.5, we derive from (3.33) and (3.34) that

$$\begin{aligned} &\mathbb{E}\Big(\sup_{-\tau\leq s\leq t} |Y(s)|^p\Big) \\ &\lesssim_T \|\xi\|_\infty^p + \mathbb{E}\Big(\int_0^t |b(\overline{Y}_{s_\delta})|^2 \mathrm{d}s\Big)^{p/2} + \mathbb{E}\Big(\int_0^t \|\sigma(\overline{Y}_{s_\delta})\|_{HS}^2 \mathrm{d}s\Big)^{p/2} \\ &\quad + \mathbb{E}\Big(\int_0^t \int_\Gamma |\gamma(\overline{Y}_{s_\delta}, u)|^2 \nu(\mathrm{d}u)\mathrm{d}s\Big)^{p/2} + \int_0^t \int_\Gamma \mathbb{E}|\gamma(\overline{Y}_{s_\delta}, u)|^p \nu(\mathrm{d}u)\mathrm{d}s \\ &\lesssim_T 1 + \|\xi\|_\infty^p + \int_0^t \mathbb{E}\Big(\sup_{-\tau\leq r\leq s} |Y(r)|^p\Big)\mathrm{d}s. \end{aligned}$$

As a result, the desired assertion follows from the Gronwall inequality.

Lemma 3.12 *Under* (A3) *and* (H1)–(H2),

$$\mathbb{E}|Y(t+\theta) - \overline{Y}_t(\theta)|^p \lesssim_T \delta \tag{3.35}$$

for any $p \geq 2$, $\theta \in [-\tau, 0]$ *and* $t \in [0, T]$.

Proof For notation simplicity, set $k_\delta := t_\delta + \theta_\delta$. By (3.31), one has

$$\overline{Y}_{t_\delta}(\theta) = \overline{Y}(k_\delta) + \frac{\theta - \theta_\delta}{\delta}\{\overline{Y}(k_\delta + \delta) - \overline{Y}(k_\delta)\}, \quad \theta \in [-\tau, 0],$$

which further implies

$$\mathbb{E}|Y(t+\theta) - \overline{Y}_t(\theta)|^p \lesssim \mathbb{E}|\overline{Y}(k_\delta + \delta) - \overline{Y}(k_\delta)|^p + \mathbb{E}|Y(t+\theta) - \overline{Y}(k_\delta)|^p. \quad (3.36)$$

For any $k_\delta \geq 0$, by Hölder's inequality [99, Theorem 7.1, p. 39] and Lemma 3.5, we obtain from (3.30) and (3.34) that

$$\begin{aligned}
&\mathbb{E}|\overline{Y}(k_\delta + \delta) - \overline{Y}(k_\delta)|^p \\
&\lesssim \delta^{p/2}(\mathbb{E}|b(\overline{Y}_{k_\delta})|^p + \mathbb{E}\|\sigma(\overline{Y}_{k_\delta})\|_{HS}^p) + \int_{k_\delta}^{k_\delta+\delta}\int_\Gamma \mathbb{E}|\gamma(\overline{Y}_{k_\delta}, u)|^p \nu(\mathrm{d}u)\mathrm{d}s \\
&\quad + \mathbb{E}\Big(\int_{k_\delta}^{k_\delta+\delta}\int_\Gamma |\gamma(\overline{Y}_{k_\delta}, u)|^2 \nu(\mathrm{d}u)\mathrm{d}s\Big)^{p/2} \qquad (3.37)\\
&\lesssim \delta\Big(\mathbb{E}|b(\overline{Y}_{k_\delta})|^p + \mathbb{E}\|\sigma(\overline{Y}_{k_\delta})\|_{HS}^p + \mathbb{E}\Big(\int_\Gamma |\gamma(\overline{Y}_{k_\delta}, u)|^2 \nu(\mathrm{d}u)\Big)^{p/2} \\
&\quad + \int_\Gamma \mathbb{E}|\gamma(\overline{Y}_{k_\delta}, u)|^p \nu(\mathrm{d}u)\Big) \\
&\lesssim \delta(1 + \mathbb{E}\|\overline{Y}_{k_\delta}\|_\infty^p) \\
&\lesssim_T \delta,
\end{aligned}$$

where in the last step we have used Lemma 3.11. In what follows, we divide the following five cases to show (3.35).

Case 1: $k_\delta \geq 0$ and $0 \leq t + \theta - k_\delta < \delta$. (3.35) holds by carrying out the argument to derive (3.37).

Case 2: $k_\delta \geq 0$ and $\delta \leq t + \theta - k_\delta < 2\delta$. It is easily seen that

$$\mathbb{E}|Y(t+\theta) - \overline{Y}(k_\delta)|^p \lesssim \mathbb{E}|Y(t+\theta) - \overline{Y}(k_\delta + \delta)|^p + \mathbb{E}|\overline{Y}(k_\delta + \delta) - \overline{Y}(k_\delta)|^p.$$

This, by utilizing (3.37) and taking Case 1 into consideration, leads to (3.35).

Case 3: $k_\delta = -\delta$ and $0 \leq t + \theta - k_\delta \leq \delta$. In this case, $-\delta \leq t + \theta \leq 0$. Thus, (3.35) follows from (A3).

Case 4: $k_\delta = -\delta$ and $\delta \leq t + \theta - k_\delta < 2\delta$. In such case, $0 \leq t + \theta < \delta$. Combining Case 1 with Case 2 gives (3.35).

Case 5: $k_\delta \leq -2\delta$. In this case, $t + \theta < 0$. So (A3) yields (3.35).

Consequently, (3.35) follows by taking Cases 1–5 into account.

The following theorem reveals convergence rate of the EM scheme (3.30) under the global Lipschitz condition, and plays a crucial role in establishing convergence rate of the EM scheme associated with (3.29) under a logarithmic Lipschitz condition below.

Theorem 3.13 *Under* (A3) *and* (H1)–(H2),

$$\mathbb{E}\Big(\sup_{0\le t\le T}|X(t)-Y(t)|^p\Big)\le c_1(L^{p/2}+K^{(p)})e^{c_2(L^{p/2}+K^{(p)})}\delta,\quad p\ge 2 \tag{3.38}$$

for some constants $c_1, c_2>0$, independent of δ.

Proof By Hölder's inequality, B–D–G's inequality, and Lemma 3.5,

$$\begin{aligned}
&\mathbb{E}\Big(\sup_{0\le s\le t}|X(t)-Y(t)|^p\Big)\\
&\lesssim \int_0^t\{\mathbb{E}|b(X_s)-b(\overline{Y}_{s_\delta})|^p+\mathbb{E}\|\sigma(X_s)-\sigma(\overline{Y}_{s_\delta})\|_{HS}^p\}\mathrm{d}s\\
&\quad+\int_0^t\int_\Gamma \mathbb{E}|\gamma(X_s,u)-\gamma(\overline{Y}_{s_\delta},u)|^p\nu(\mathrm{d}u)\mathrm{d}s\\
&\quad+\mathbb{E}\Big(\int_0^t\int_\Gamma \mathbb{E}|\gamma(X_s,u)-\gamma(\overline{Y}_{s_\delta},u)|^2\nu(\mathrm{d}u)ds\Big)^{p/2}\\
&\lesssim \int_0^t\{\mathbb{E}|b(X_s)-b(Y_s)|^p+\mathbb{E}\|\sigma(X_s)-\sigma(Y_s)\|_{HS}^p\}\mathrm{d}s \qquad (3.39)\\
&\quad+\int_0^t\{\mathbb{E}|b(Y_s)-b(\overline{Y}_{s_\delta})|^p+\mathbb{E}\|\sigma(Y_s)-\sigma(\overline{Y}_{s_\delta})\|_{HS}^p\}\mathrm{d}s\\
&\quad+\int_0^t\int_\Gamma \mathbb{E}|\gamma(X_s,u)-\gamma(Y_s,u)|^p\nu(\mathrm{d}u)\mathrm{d}s\\
&\quad+\int_0^t\int_\Gamma \mathbb{E}|\gamma(Y_s,u)-\gamma(\overline{Y}_{s_\delta},u)|^p\nu(\mathrm{d}u)\mathrm{d}s,\\
&\quad+\mathbb{E}\Big(\int_0^t\int_\Gamma \mathbb{E}|\gamma(X_s,u)-\gamma(Y_s,u)|^2\nu(\mathrm{d}u)ds\Big)^{p/2}\\
&\quad+\mathbb{E}\Big(\int_0^t\int_\Gamma \mathbb{E}|\gamma(Y_s,u)-\gamma(\overline{Y}_{s_\delta},u)|^2\nu(\mathrm{d}u)ds\Big)^{p/2},\quad t\in[0,T].
\end{aligned}$$

In view of (H1) and (H2), in addition to Hölder's inequality and (3.35),

$$\begin{aligned}
\text{RHS of (3.39)}&\lesssim (L^{p/2}+K^{(p)})\int_0^t\Big\{\mathbb{E}\Big(\sup_{0\le r\le s}|X(r)-Y(r)|^p\Big)\\
&\quad+\int_{[-\tau,0]}\mathbb{E}|Y(s+\theta)-\bar{Y}_s(\theta)|^p\mu(\mathrm{d}\theta)\Big\}\mathrm{d}s\\
&\lesssim (L^{p/2}+K^{(p)})\Big\{\delta+\int_0^t\mathbb{E}\Big(\sup_{0\le r\le s}|X(r)-Y(r)|^p\Big)\mathrm{d}s\Big\}.
\end{aligned}$$

Substituting this into (3.39) and applying Gronwall's inequality yields the desired assertion.

Remark 3.14 In (3.29) with $\gamma \equiv 0$, Theorem 3.13 holds with $K^{(p)} = 0$. However, for such case, condition (H1) can be strengthened as

$$|b(\xi) - b(\eta)|^2 \vee \|\sigma(\xi) - \sigma(\eta)\|_{HS}^2 \le L\|\xi - \eta\|_\infty^2$$

for some $L > 0$.

With Theorem 3.13 in mind, in what follows we shall discuss convergence rate of the EM scheme associated with (3.29) under a logarithm Lipschitz condition. We assume the following conditions hold.

(H4) For each $j \in \mathbb{N}$ and $\xi, \eta \in \mathscr{D}$ with $\max\{\|\xi\|_\infty, \|\eta\|_\infty\} \le j$, there exists an $L_j \ge 1$ such that

$$|b(\xi) - b(\eta)|^2 \vee \|\sigma(\xi) - \sigma(\eta)\|_{HS}^2 \le L_j \int_{[-\tau,0]} |\varphi(\theta) - \psi(\theta)|^2 \mu(\mathrm{d}\theta),$$

and, for any $p > 2$, there exists $K_j^{(p)} > 0$ such that

$$\Big(\int_\Gamma |\gamma(\xi, u) - \gamma(\eta, u)|^2 \nu(\mathrm{d}u)\Big)^{p/2} + \int_\Gamma |\gamma(\xi, u) - \gamma(\eta, u)|^p \nu(\mathrm{d}u)$$
$$\le K_j^{(p)} \int_{[-\tau,0]} |\varphi(\theta) - \psi(\theta)|^p \mu(\mathrm{d}\theta).$$

(H5) For any $\xi \in \mathscr{D}$ and $p > 2$, there exist $K_0, K_p > 0$ such that

$$|b(\xi)|^2 \vee \|\sigma(\xi)\|_{HS}^2 \le K_0(1 + \|\xi\|_\infty^2), \quad \text{and}$$
$$\Big(\int_\Gamma |\gamma(\xi, u)|^2 \nu(\mathrm{d}u)\Big)^{p/2} + \int_\Gamma |\gamma(\xi, u)|^p \nu(\mathrm{d}u) \le K_p(1 + \|\xi\|_\infty^p).$$

Our main result in this section is stated as follows.

Theorem 3.15 *Let* (A3) *and* (H4)–(H5) *hold, and assume that there exists* $\alpha = \alpha(p) > 0$ *such that*

$$L_j^{p/2} + K_j^{(p)} \le \alpha \log j. \tag{3.40}$$

Suppose further $K_j^{(p)} \ge K_j^{(2+\varepsilon)}$ *for* $\varepsilon \in (0, p-2)$. *Then,*

$$\mathbb{E}\Big(\sup_{0 \le t \le T} |X(t) - Y(t)|^2\Big) \lesssim_T \delta^{\frac{2}{2+\varepsilon}}, \quad \varepsilon \in (0, p-2). \tag{3.41}$$

Proof For each integer $j \geq 1$, set $B_j(0) := \{x \in \mathbb{R}^n : |x| \leq j\}$. Define the projection operator $\pi_j : \mathbb{R}^n \mapsto B_j(0)$ by

$$\pi_j(x) = \frac{j \wedge |x|}{|x|} x,$$

where we set $\pi_j(0) = 0$ as usual. It is easy to see that

$$|\pi_j(x) - \pi_j(y)| \leq |x - y|, \quad x, y \in \mathbb{R}^n. \tag{3.42}$$

Define the operator $\overline{\pi}_j : \mathscr{D} \mapsto \mathscr{D}$ by

$$\overline{\pi}_j(\varphi) = \{\pi_j(\varphi(\theta)) : -\tau \leq \theta \leq 0\}.$$

Clearly,

$$\|\overline{\pi}_j(\varphi)\|_\infty \leq j, \quad \varphi \in \mathscr{D}. \tag{3.43}$$

Define the truncation mappings $b_j : \mathscr{D} \mapsto \mathbb{R}^n$, $\sigma_j : \mathscr{D} \mapsto \mathbb{R}^n \otimes \mathbb{R}^m$ and $\gamma_j : \mathscr{D} \times \Gamma \mapsto \mathbb{R}^n$ by

$$b_j(\varphi) = b(\overline{\pi}_j(\varphi)), \;\; \sigma_j(\varphi) = \sigma(\overline{\pi}_j(\varphi)), \;\; \gamma_j(\varphi, u) = \gamma(\overline{\pi}_j(\varphi), u), \quad u \in \Gamma.$$

For $t \in [0, T]$, let $X^{(j)}(t)$ solve the following FSDE

$$\mathrm{d}X^{(j)}(t) = b_j(X_t^{(j)})\mathrm{d}t + \sigma_j(X_t^{(j)})\mathrm{d}W(t) + \int_\Gamma \gamma_j(X_t^{(j)}, u)\widetilde{N}(\mathrm{d}t, \mathrm{d}u), \quad t > 0 \tag{3.44}$$

with the initial data $X_0^{(j)} = \xi$. In what follows, we stipulate $\varepsilon \in (0, p-2)$. For any $\varphi, \psi \in \mathscr{D}$, according to (H3), (3.42), and (3.43), it follows that

$$\begin{aligned}|b_j(\varphi) - b_j(\psi)|^2 \vee |\sigma_j(\varphi) - \sigma_j(\psi)|^2 &\leq L_j \int_{-\tau}^0 |\pi_j(\varphi(\theta)) - \pi_j(\psi(\theta))|^2 \mu(\mathrm{d}\theta)\\ &\leq L_j \int_{-\tau}^0 |\varphi(\theta) - \psi(\theta)|^2 \mu(\mathrm{d}\theta),\end{aligned}$$

and that

$$\int_\Gamma |\gamma_j(\varphi, u) - \gamma_j(\psi, u)|^{2+\varepsilon} \nu(\mathrm{d}u) \leq K_j^{(2+\varepsilon)} \int_{[-\tau,0]} |\varphi(\theta) - \psi(\theta)|^{2+\varepsilon} \mu(\mathrm{d}\theta).$$

Thus, by Theorem 3.13, one has

$$\mathbb{E}\Big(\sup_{0 \leq t \leq T} |X^{(j)}(t) - Y^{(j)}(t)|^{2+\varepsilon}\Big) \leq c_1(L_j^{1+\varepsilon/2} + K_j^{(2+\varepsilon)})e^{c_2(L_j^{1+\varepsilon/2} + K_j^{(2+\varepsilon)})}\delta$$

for some constants $c_1, c_2 > 0$, where $Y^{(j)}(t)$ is the corresponding continuous-time EM scheme associated with (3.44). This, together with (3.40) and $K_j^{(p)} \geq K_j^{(2+\varepsilon)}$ with $\varepsilon \in (0, p-2)$, gives that

$$\mathbb{E}\Big(\sup_{0\leq t\leq T} |X^{(j)}(t) - Y^{(j)}(t)|^{2+\varepsilon}\Big) \leq e^{(c_1+c_2)(L_j^{p/2}+K_j^{(p)})}\delta \leq j^{\alpha(c_1+c_2)}\delta. \tag{3.45}$$

For any integer $j \geq 1$, define the stopping time

$$\tau_j = T \wedge \inf\{t \in [0, T] : \|X_t^{(j)}\|_\infty \vee \|Y_t^{(j)}\|_\infty \geq j\}.$$

It is trivial to see that $\|X_{s-}^{(j)}\|_\infty \leq j$ for any $0 \leq s < \tau_j$. For any $0 \leq s < \tau_j$, note that

$$b_j(X_{s-}^{(j)}) = b_{j+1}(X_{s-}^{(j)}) = b(X_{s-}^{(j)}),$$

and

$$\sigma_j(X_{s-}^{(j)}) = \sigma_{j+1}(X_{s-}^{(j)}) = \sigma(X_{s-}^{(j)}), \quad \gamma_j(X_{s-}^{(j)}, u) = \gamma_{j+1}(X_{s-}^{(j)}, u) = \gamma(X_{s-}^{(j)}, u).$$

Consequently, we have

$$X(t) = X^{(j)}(t) = X^{(j+1)}(t), \quad 0 \leq t < \tau_j. \tag{3.46}$$

Also, we derive that

$$Y(t) = Y^{(j)}(t) = Y^{(j+1)}(t), \quad 0 \leq t < \tau_j.$$

Therefore, τ_j is nondecreasing and then $\lim_{j\to\infty} \tau_j = T$ a.s. Observe from (3.46) that

$$\begin{aligned}|X(t) - Y(t)|^2 &= \sum_{j=1}^{\infty} |X(t) - Y(t)|^2 I_{[\tau_{j-1}, \tau_j)}(t)\\ &= \sum_{j=1}^{\infty} |X^{(j)}(t) - Y^{(j)}(t)|^2 I_{[\tau_{j-1}, \tau_j)}(t)\\ &\leq \sum_{j=1}^{\infty} |X^{(j)}(t) - Y^{(j)}(t)|^2 I_{\{j-1\leq \|X\|_T \vee \|Y\|_T\}},\end{aligned}$$

where $\|X\|_T := \sup_{0\leq t\leq T} |X(t)|$ and $\|Y\|_T := \sup_{0\leq t\leq T} |Y(t)|$. Therefore, by the Hölder inequality,

$$\mathbb{E}\Big(\sup_{0\le t\le T}|X(t)-Y(t)|^2\Big)\le\sum_{j=1}^{\infty}\Big(\mathbb{E}\Big(\sup_{0\le t\le T}|X^{(j)}(t)-Y^{(j)}(t)|^{2+\varepsilon}\Big)\Big)^{\frac{2}{2+\varepsilon}}$$
$$\times\,(\mathbb{P}(j-1\le\|X\|_T\vee\|Y\|_T)^{\frac{\varepsilon}{2+\varepsilon}}. \tag{3.47}$$

Next, for any $q\ge 2$, we obtain from Lemma 3.11 and Chebyshev's inequality that

$$\mathbb{P}(j-1\le\|X\|_T\vee\|Y\|_T)\le\frac{\mathbb{E}\|X\|_T^q+\mathbb{E}\|Y\|_T^q}{(\frac{j}{2})^q}\lesssim j^{-q},\quad j\ge 2. \tag{3.48}$$

Substituting (3.45) and (3.48) into (3.47), one has

$$\mathbb{E}\Big(\sup_{0\le t\le T}|X(t)-Y(t)|^2\Big)\lesssim\Big(1+\sum_{j=2}^{\infty}j^{\frac{2\alpha(c_1+c_2)-q\varepsilon}{2+\varepsilon}}\Big)\delta^{\frac{2}{2+\varepsilon}}. \tag{3.49}$$

Taking $q>0$ such that

$$q>\varepsilon^{-1}(2\alpha(c_1+c_2)+2+\varepsilon),$$

we see that the series on the right-hand side of (3.49) is convergent, whence the desired assertion (3.41) follows.

Chapter 4
Large Deviations for FSDEs

In this chapter, by the weak convergence method, based on a variational representation for positive functionals of a Poisson random measure and a Brownian motion, we establish uniform large deviation principles (LDPs for short) for a class of FSDEs of neutral type driven by jump processes. As a byproduct, we also obtain uniform LDPs for SDDEs of neutral type, which in particular, allows the coefficients to be highly nonlinear w.r.t. the delay variables. Moreover, we obtain a moderate deviations principle (MDP for abbreviation) for FSDEs with jumps, and a central limit theorem (CLT) for FSDEs driven by Brownian motions. Sections 4.1–4.4 of this chapter are based on [17].

4.1 Introduction

As is well known, LDP is concerned with the asymptotic behavior of tail probability estimates for rare events on an exponential scale; see, for example, Dembo–Zeitouni [44] and Freidlin–Wentzell [58]. As stated in Maroulas [104], the theory of LDPs is applicable to many fields such as statistical inference, queuing systems, communication networks, risk-sensitive control, and statistic mechanics; see [44, 45, 51, 134] and references therein. The classical method for establishing LDPs relies on some discretization/approximation arguments and exponential-type probability estimates; see e.g., [57, 110]. However, this method may be cumbersome, in particular, for SPDEs and FSDEs. In recent years, the weak convergence approach due to Budhiraja–Dupuis [23], based on a variational representation of bounded continuous functionals of Brownian motions, has been applied in establishing LDPs for a wide range of stochastic dynamical systems driven by Brownian motions [120, 156] for SDEs and [97, 125] for SPDEs. In contrast with the discretization method, the appealing advantage of weak convergence approach is that one can bypass the approximation arguments and avoid exponential-type probability estimates. Although there are extensive results on LDPs for stochastic dynamical systems driven by finite- and

J. Bao et al., *Asymptotic Analysis for Functional Stochastic Differential Equations*, SpringerBriefs in Mathematics, DOI 10.1007/978-3-319-46979-9_4

infinite-dimensional Brownian motions, there is not much work on the topic for stochastic models driven by jump processes. de Acosta [39] studied LDPs for SDEs with bounded coefficients driven by Poisson measures; Hult–Samorodnitsky [70] investigated LDPs for point processes with heavy tails; Röckner–Zhang [124] obtained LDPs for a class of infinite-dimensional Ornstein–Uhlenbeck type processes with jumps. For an SPDE with multiplicative Lévy noise, we refer to Sowers–Zabczyk [132]. Recently, Budhiraja et al. [26] developed a variational representation for positive functionals of a Poisson random measure and a Brownian motion, which is analogous to the one established in [23] for the Brownian motion case. This new variational representation has also been applied effectively to establish LDPs for stochastic equations with jumps; see e.g., [26, 146] for SDEs and [24, 104, 149] for SPDEs. In particular, Maroulas [104], Wu [146], and Zhang [156] studied the uniform LDPs. In general, the uniformity is w.r.t. the initial value ξ, taking values in some compact subset of a Polish space, of the stochastic equations involved. For stochastic systems with memory, there are only a few results on LDPs. By adopting time-discretization arguments, Scheutzow [127] and Mohammed-Zhang [110] established large deviation results for SDDEs with additive and multiplicative Brownian motion noises, respectively.

In what follows, Section 4.2 focuses on collecting some notation and recalls a general criteria on the uniform LDPs. Sections 4.3 and 4.4 are devoted to the uniform large deviations for FSDEs of neutral type under uniform Lipschitz conditions and SDDEs of neutral type under polynomial growth condition, respectively. Section 4.5 is concerned with MDPs for FSDEs with jumps. In the last section, we obtain a central limit theorem (CLT) for a class of FSDEs driven by Brownian motions.

4.2 Preliminaries

Let $\mathbb{S}_0$ and $\mathbb{S}$ be Polish spaces (i.e., complete separable metrizable topological spaces) and $\mathscr{B}(\mathbb{S})$ the Borel σ-algebra generated by all open sets in $\mathbb{S}$. For a closed interval $\Lambda \subset (-\infty, \infty)$, denoted by $C(\Lambda; \mathbb{S})$ (resp., $D(\Lambda; \mathbb{S})$) the family of all continuous (resp., càdlàg) functions $f : \Lambda \mapsto \mathbb{S}$ endowed with the uniform topology (resp., the Skorokhod topology). Let $M_b(\mathbb{S})$ be the collection of all bounded $\mathscr{B}(\mathbb{S})/\mathscr{B}(\mathbb{R})$-measurable maps and $C_b(\mathbb{S})$ means the set of all bounded continuous functions $f : \mathbb{S} \mapsto \mathbb{R}$. Denote by $C_c(\mathbb{S})$ the space of all continuous functions $f : \mathbb{S} \mapsto \mathbb{R}$ with compact support. The symbol "$\Rightarrow$" means convergence in distribution of random variables. For each $\xi \in \mathbb{S}_0$, let $(X^\varepsilon(\xi))_{\varepsilon>0}$ be a family of all $\mathbb{S}$-valued random variables defined on a probability space $(\Omega, \mathscr{F}, \mathbb{P})$. The theory of large deviations concerns with exponential decay scale for the law of $(X^\varepsilon(\xi))_{\varepsilon>0}$ as $\varepsilon \to 0$. The exponential decay rate is typically expressed in terms of "rate function" below.

Definition 4.1 A function $I : \mathbb{S} \mapsto [0, \infty]$ is called a rate function, if, for each $a < \infty$, the level set $\{f \in \mathbb{S} : I(f) \leq a\}$ is a compact subset of $\mathbb{S}$. A family of rate functions $I_\xi : \mathbb{S} \mapsto [0, \infty]$, parameterized by $\xi \in \mathbb{S}_0$, is said to have compact level sets on compacts if, for all compact subsets $K \subset \mathbb{S}_0$ and each $M < \infty$, $\Lambda_{M,K} := \bigcup_{\xi \in K}\{f \in \mathbb{S} : I_\xi(f) \leq M\}$ is a compact subset of $\mathbb{S}$.

Definition 4.2 The sequence $(X^\varepsilon(\xi))_{\epsilon>0}$ is said to satisfy a uniform LDP with rate function I_ξ, if, for each $A \in \mathscr{B}(\mathbb{S})$ and compact subset $K \subset \mathbb{S}_0$,

$$\begin{aligned} -\inf_{f\in A^\circ} I_\xi(f) &\leq \liminf_{\varepsilon\to 0} \varepsilon \log \mathbb{P}(X^\varepsilon(\xi) \in A) \\ &\leq \limsup_{\varepsilon\to 0} \varepsilon \log \mathbb{P}(X^\varepsilon(\xi) \in A) \\ &\leq -\inf_{f\in \overline{A}} I_\xi(f), \qquad \xi \in K, \end{aligned}$$

where A° and $\overline{A}$ denote the interior and closure of A, respectively.

According to [51, Theorem 1.2.1, p. 4] and [51, Theorem 1.2.3, p. 7], the uniform LDP is equivalent to the uniform Laplace principle below.

Definition 4.3 Let $I_\xi : \mathbb{S} \mapsto [0, \infty]$, parameterized by $\xi \in \mathbb{S}_0$, be a family of rate functions with compact level sets. The sequence $(X^\varepsilon(\xi))_{\varepsilon>0}$ is said to satisfy a uniform Laplace principle with the rate function I_ξ, if for each compact subset $K \subset \mathbb{S}_0$ and $g \in C_b(\mathbb{S})$,

$$\limsup_{\varepsilon\to 0} \sup_{\xi\in K} \left| \varepsilon \log \mathbb{E}_\xi \Big(\exp\Big[-\frac{g(X^\varepsilon(\xi))}{\varepsilon} \Big] \Big) + \inf_{f\in\mathbb{S}} \{ g(f) + I_\xi(f) \} \right| = 0.$$

To present the general criteria on uniform LDPs for stochastic dynamical systems driven by jump processes, we further need to recall some additional notation and notions; see [26] for more details. For a locally compact Polish space $(\mathbb{X}, \|\cdot\|_{\mathbb{X}})$ (i.e., every point of $\mathbb{X}$ has a compact neighborhood), denote by $\mathscr{M}(\mathbb{X})$ the family of all locally finite measures ν on $(\mathbb{X}, \mathscr{B}(\mathbb{X}))$ (i.e., $\nu(K) < \infty$ for every compact subset $K \subset \mathbb{X}$). Endow $\mathscr{M}(\mathbb{X})$ with the weakest topology such that for every $f \in C_c(\mathbb{X})$, the function $\nu \mapsto \langle f, \nu \rangle := \int_{\mathbb{X}} f(u)\nu(du)$, $\nu \in \mathbb{M}(\mathbb{X})$, is continuous. This topology can be metrized such that $\mathscr{M}(\mathbb{X})$ is a Polish space; see e.g., [26]. Throughout this chapter, let $T > 0$ and $\nu \in \mathscr{M}(\mathbb{X})$ be fixed. Let $\mathbb{X}_T = [0, T] \times \mathbb{X}$, $\mathbb{M} = \mathscr{M}(\mathbb{X}_T)$, $\mathbb{W} = C([0, T]; \mathbb{R}^m)$, and $\mathbb{V} = \mathbb{W} \times \mathbb{M}$. Let $\nu_T = \lambda_T \otimes \nu$, where λ_T is the Lebesgue measure on $[0, T]$. For $\theta > 0$, let $\mathbb{P}_\theta$ be the unique probability measure on $(\mathbb{V}, \mathscr{B}(\mathbb{V}))$ such that the following conditions hold.

(i) The canonical map $W : \mathbb{V} \mapsto \mathbb{W}$, $W(w, m) := w$ is an m-dimensional standard Brownian motion;
(ii) The canonical map $N : \mathbb{V} \mapsto \mathbb{M}$, $N(w, m) := m$ is a Poisson random measure with intensity measure $\theta\nu_T$;
(iii) $(W(t))_{t\in[0,T]}$ and $(N((0, t] \times A) - \theta t \nu(A))_{t\in[0,T]}$, $A \in \mathscr{B}(\mathbb{X})$, are $\mathscr{G}_t$-martingales, where

$$\mathscr{G}_t := \sigma\{N((0, s] \times A), W(s) : s \in (0, t], A \in \mathscr{B}(\mathbb{X})\}.$$

The corresponding expectation operator will be denoted as $\mathbb{E}_\theta$.

Let $\mathbb{Y} = \mathbb{X} \times [0, \infty)$, $\mathbb{Y}_T = [0, T] \times \mathbb{Y}$, $\overline{\mathbb{M}} = \mathscr{M}(\mathbb{Y}_T)$, and $\overline{\mathbb{V}} = \mathbb{W} \times \overline{\mathbb{M}}$. Define $(\overline{\mathbb{P}}, (\overline{\mathscr{G}}_t))$ on $(\overline{\mathbb{V}}, \mathscr{B}(\overline{\mathbb{V}}))$ analogous to $(\mathbb{P}_\theta, \mathscr{G}_t)$ by replacing $(N, \theta\nu_T)$ with $(\overline{N}, \overline{\nu}_T)$ in the above with $\overline{\nu}_T := \lambda_T \otimes \nu \otimes \lambda_\infty$, where λ_∞ is the Lebesgue measure on $[0, \infty)$. Let

$(\overline{\mathscr{F}}_t)$ be the $\overline{\mathbb{P}}$-completion of $(\overline{\mathscr{G}}_t)$ and $\overline{\mathscr{P}}$ the predictable σ-field on $[0, T] \times \overline{\mathbb{V}}$ with the filtration $(\overline{\mathscr{F}}_t)_{t\in[0,T]}$ on $(\overline{\mathbb{V}}, \mathscr{B}(\overline{\mathbb{V}}))$. Define the Cameron–Martin space $\overline{\mathscr{A}}_1$ by

$$\overline{\mathscr{A}}_1 = \Big\{h : [0, T] \mapsto \mathbb{R}^m \Big| \; h \text{ is } \overline{\mathscr{P}}/\mathscr{B}(\mathbb{R}^m)\text{-measurable}, \tag{4.1}$$
$$h(t) = \int_0^t \dot{h}(s)\mathrm{d}s, \; t \in [0, T], \text{ and } \int_0^T |\dot{h}(s)|^2 \mathrm{d}s < \infty, \quad \overline{\mathbb{P}} - \text{a.s.}\Big\},$$

where the dot denotes the generalized derivative, and

$$\overline{\mathscr{A}}_2 := \{\varphi : \mathbb{X}_T \times \overline{\mathbb{V}} \mapsto [0, \infty) | \varphi \text{ is } (\overline{\mathscr{P}} \otimes \mathscr{B}(\mathbb{X}))/\mathscr{B}([0, \infty))\text{-measurable}\}. \tag{4.2}$$

For $\varphi \in \overline{\mathscr{A}}_2$, define a counting measure N^φ on $\mathbb{X}_T$ by

$$N^\varphi((0, t] \times U) = \int_0^t \int_U \int_0^\infty 1_{[0,\varphi(s,x)]}(r)\overline{N}(\mathrm{d}s, \mathrm{d}x, \mathrm{d}r), \quad t \in [0, T], \; U \in \mathscr{B}(\mathbb{X}). \tag{4.3}$$

Here, N^φ is called a controlled random counting measure with $\varphi(s, x)$ selecting the intensity for the points at time s and location x. Moreover, if $\varphi(s, x, \omega) \equiv \theta$, $\forall\, (s, x, \omega) \in \mathbb{X}_T \times \overline{\mathbb{V}}$, for some $\theta > 0$, we write N^θ instead of N^φ. Define $l : [0, \infty) \mapsto [0, \infty)$ by

$$l(r) = r \log r - r + 1, \quad r \in (0, \infty).$$

Set $\mathscr{U} := \overline{\mathscr{A}}_1 \times \overline{\mathscr{A}}_2$ and

$$\overline{L}_T(u) := L_{1,T}(h) + L_{2,T}(\varphi), \qquad u = (h, \varphi) \in \mathscr{U}, \tag{4.4}$$

where

$$L_{1,T}(h) := \frac{1}{2}\int_0^T |\dot{h}(t)|^2 \mathrm{d}t \;\text{ and }\; L_{2,T}(\varphi) := \int_{\mathbb{X}_T} l(\varphi(t, x, \omega))\nu_T(\mathrm{d}t, \mathrm{d}x). \tag{4.5}$$

Remark 4.4 Budhiraja et al. [26, Theorem 3.1] established the following variational formula: for $F \in M_b(\mathbb{V})$ and $\theta > 0$,

$$-\log \mathbb{E}_\theta(e^{F(W,N)}) = -\log \overline{\mathbb{E}} e^{F(W,N^\theta)} = \inf_{u=(\psi,\varphi)\in\mathscr{U}} \overline{\mathbb{E}}(\theta \overline{L}_T(u) + F(W^{\theta\psi}, N^{\theta\varphi})),$$

where $W^\psi(t) := W(t) + \int_0^t \dot{\psi}(s)\mathrm{d}s$ for $\psi \in \overline{\mathscr{A}}_1$. This representation leads to the equivalence between the LDP and the Laplace principle in a Polish space; see e.g., [23, Theorem 4.4].

For each $N > 0$, let

$$S_{1,N} = \{h : [0,T] \mapsto \mathbb{R}^m : L_{1,T}(h) \le N\} \text{ and } S_{2,N} = \{\varphi : \mathbb{X}_T \mapsto [0,\infty) : L_{2,T}(\varphi) \le N\}.$$

The function $\varphi \in S_{2,N}$ can be identified with a measure $\nu_T^\varphi \in \mathbb{M}(\mathbb{X}_T)$, defined as

$$\nu_T^\varphi(A) = \int_A \varphi(s,x)\nu_T(\mathrm{d}s,\mathrm{d}x), \quad A \in \mathscr{B}(\mathbb{X}_T).$$

This identification induces a topology on $S_{2,N}$ under which $S_{2,N}$ is a compact space [24, Theorem A.1]. In what follows, we use this topology on $S_{2,N}$. Let $S_N = S_{1,N} \times S_{2,N}$, $S = \bigcup_{N\ge 1} S_N$ and

$$\mathscr{U}_N = \{u = (h,\varphi) \in \mathscr{U} : u(\omega) \in S_N, \ \overline{\mathbb{P}} - \text{a.s.}\}.$$

Let $\{K_n \subset \mathbb{X}, n = 1,2\ldots\}$ be an increasing sequence of compact sets such that $\bigcup_{n=1}^\infty K_n = \mathbb{X}$. For each n, let

$$\overline{\mathscr{A}}_{b,n} = \{\varphi \in \overline{\mathscr{A}}_2 : \text{ for all } (t,\omega) \in [0,T] \times \overline{\mathbb{M}}, 1/n \le \varphi(t,x,\omega) \le n \text{ if } x \in K_n \\ \text{and } \varphi(t,x,\omega) = 1 \text{ if } x \in K_n^c\},$$

$\overline{\mathscr{A}}_b = \bigcup_{n=1}^\infty \overline{\mathscr{A}}_{b,n}$, and $\overline{\mathscr{U}}_N = \mathscr{U}_N \bigcap \{(h,\varphi) : \varphi \in \overline{\mathscr{A}}_b\}$.

Let $K \subset \mathbb{S}_0$ be a compact subset and $(\mathscr{G}^\varepsilon)_{\varepsilon>0}$ a family of measurable maps from $K \times \bar{\mathbb{V}}$ to $\mathbb{S}$. Set

$$X^\varepsilon(\cdot;\xi) := \mathscr{G}^\varepsilon(\xi, \sqrt{\varepsilon}W, \varepsilon N^{\varepsilon^{-1}}). \tag{4.6}$$

Fix an $N \in \mathbb{N}$ and assume that there exists a measurable mapping $\mathscr{G}^0 : \mathbb{S}_0 \times \mathbb{V} \mapsto \mathbb{S}$ such that the following holds:

(i) For $(h_n,\varphi_n), (h,\varphi) \in S_N$, and $\xi_n, \xi \in K$ such that $(\xi_n, h_n, \varphi_n) \to (\xi, h, \varphi)$ as $n \to \infty$,

$$\mathscr{G}^0\Big(\xi_n, \int_0^\cdot \dot{h}_n(s)\mathrm{d}s, \nu_T^{\varphi_n}\Big) \to \mathscr{G}^0\Big(\xi, \int_0^\cdot \dot{h}(s)\mathrm{d}s, \nu_T^{\varphi}\Big);$$

(ii) For $u_\varepsilon := (h_\varepsilon, \varphi_\varepsilon), u := (h,\varphi) \in \bar{\mathscr{U}}_N$ and $\xi_\varepsilon, \xi \in K$ such that $u_\varepsilon \Rightarrow u$ and $\xi_\varepsilon \to \xi$ as $\varepsilon \to 0$,

$$\mathscr{G}^\varepsilon\Big(\xi_\varepsilon, \sqrt{\varepsilon}W + \int_0^\cdot \dot{h}_\varepsilon(s)\mathrm{d}s, \varepsilon N^{\varepsilon^{-1}\varphi_\varepsilon}\Big) \Rightarrow \mathscr{G}^0\Big(\xi, \int_0^\cdot \dot{h}(s)\mathrm{d}s, \nu_T^\varphi\Big).$$

For any $\xi \in \mathbb{S}_0$ and $f \in \mathbb{S}$, define

$$I_\xi(f) = \inf_{u=(h,\varphi)\in\mathbb{S}_f} \overline{L}_T(u), \tag{4.7}$$

where, by convention, $I_\xi(f) = \infty$ if $\mathbb{S}_f = \emptyset$, and

$$\mathbb{S}_f := \Big\{u = (h, \varphi) \in S : f = \mathscr{G}^0\Big(\xi, \int_0^\cdot \dot{h}(s)\mathrm{d}s, \nu_T^\varphi\Big)\Big\}.$$

The following uniform LDP criteria was presented in [104, Theorem 4.4].

Lemma 4.5 *Let $(X^\varepsilon(\cdot;\xi))_{\varepsilon>0}$ be defined by* (4.6) *and assume that* (i) *and* (ii) *hold. Suppose further that, for all $f \in \mathbb{S}$, $\xi \mapsto I_\xi$ is a lower semi-continuous map from $\mathbb{S}_0$ to $[0,\infty]$. Then, for all $\xi \in \mathbb{S}_0$, $f \mapsto I_\xi(f)$ is a rate function on $\mathbb{S}$ and the family $\{I_\xi(\cdot), \xi \in \mathbb{S}_0\}$ of rate functions has compact level sets on compacts. Furthermore, $(X^\varepsilon(\cdot;\xi))_{\varepsilon>0}$ satisfies the Laplace principle (hence LDP) on $\mathbb{S}$, with rate function I_ξ, uniformly on compact subsets of $\mathbb{S}_0$.*

4.3 Uniform Large Deviations under Uniform Lipschitz Conditions

In this section, we are interested in the uniform LDPs for a range of FSDEs of neutral type on $(\mathbb{R}^n, \langle\cdot,\cdot\rangle, |\cdot|)$ in the framework

$$\begin{aligned}\mathrm{d}\{X^\varepsilon(t) - G(X_t^\varepsilon)\} &= b(t, X_t^\varepsilon)\mathrm{d}t + \sqrt{\varepsilon}\sigma(t, X_t^\epsilon)\mathrm{d}W(t) \\ &\quad + \varepsilon\int_{\mathbb{X}} \phi(t, X_{t-}^\varepsilon, x)\tilde{N}^{\varepsilon^{-1}}(\mathrm{d}t, \mathrm{d}x), \quad t > 0\end{aligned} \tag{4.8}$$

with $\varepsilon > 0$, which is named as the scaling parameter, and the initial data $X_0^\epsilon = \xi \in \mathscr{C}$. Here, $G : \mathscr{D} \mapsto \mathbb{R}^n$, $b : [0,\infty) \times \mathscr{D} \mapsto \mathbb{R}^n$, $\sigma : [0,\infty) \times \mathscr{D} \mapsto \mathbb{R}^n \otimes \mathbb{R}^m$, $\phi : [0,\infty) \times \mathscr{D} \times \mathbb{X} \mapsto \mathbb{R}^n$. Moreover, $\widetilde{N}^{\varepsilon^{-1}}([0,t]\times B) := N^{\varepsilon^{-1}}([0,t]\times B) - \varepsilon^{-1}t\nu(B)$, $B \in \mathscr{B}(\mathbb{X})$, is the compensated Poisson random measure of $N^{\varepsilon^{-1}}$, which is defined in (4.3).

We further need to introduce some assumptions imposed on the coefficients of (4.8). For arbitrary $\xi, \eta \in \mathscr{D}$ and $t \in [0, T]$, we assume that

(H1) There exists $\kappa \in (0, 1)$ such that

$$|G(\xi) - G(\eta)| \le \kappa\|\xi - \eta\|_\infty;$$

(H2) There exists $\lambda = \lambda(T) > 0$ such that

$$|b(t,\xi)-b(t,\eta)|^2+\|\sigma(t,\xi)-\sigma(t,\eta)\|_{HS}^2+\int_{\mathbb{X}}|\phi(t,\xi,x)-\phi(t,\eta,x)|^2\nu(\mathrm{d}x)$$
$$\le \lambda\|\xi-\eta\|_\infty^2,\quad 0\le t\le T,$$

and

$$\rho_T:=\sup_{0\le t\le T}\left\{|b(t,0)|^2+\|\sigma(t,0)\|_{HS}^2+\int_{\mathbb{X}}|\phi(t,0,x)|^2\nu(\mathrm{d}x)\right\}<\infty.$$

Remark 4.6 By (H2), it is easily seen that there exists a $\lambda_0=\lambda_0(T)>0$ such that

$$|b(t,\xi)|^2+\|\sigma(t,\xi)\|_{HS}^2+\int_{\mathbb{X}}|\phi(t,\xi,x)|^2\nu(\mathrm{d}x)\le\lambda_0(1+\|\xi\|_\infty^2),\quad \xi\in\mathscr{D}. \tag{4.9}$$

On the other hand, under (H1) and (H2), following the line of [99, Theorem 2.2, p. 204], we conclude that (4.8) is well-posed.

For $t\in[0,T]$ and $x\in\mathbb{X}$, let

$$\Gamma_0(t,x)=\sup_{\xi\in\mathscr{D}}\frac{|\phi(t,\xi,x)|}{1+\|\xi\|_\infty}\quad\text{and}\quad\Gamma_1(t,x)=\sup_{\xi,\eta\in\mathscr{D},\xi\ne\eta}\frac{|\phi(t,\xi,x)-\phi(t,\eta,x)|}{\|\xi-\eta\|_\infty}.$$

We further assume that the following exponential integrability conditions hold for Γ_0 and Γ_1, respectively.

(H3) There exist $\delta_1,\delta_2\in(0,\infty)$ such that for all $E\in\mathscr{B}(\mathbb{X}_T)$ with $\nu_T(E)<\infty$,

$$\int_E\exp(\delta_1\Gamma_0^2(s,x))\nu_T(\mathrm{d}s,\mathrm{d}x)<\infty\quad\text{and}\quad\int_E\exp(\delta_2\Gamma_1^2(s,x))\nu_T(\mathrm{d}s,\mathrm{d}x)<\infty.$$

One of main results in this chapter is stated as follows:

Theorem 4.7 *Under* (H1)–(H3), $(X^\varepsilon(t;\xi))_{\varepsilon>0,t\in[0,T]}$ *satisfying* (4.8) *admits a uniform LDP in* $\mathbb{S}:=D([0,T];\mathbb{R}^n)$ *with rate function* $I_\xi:\mathbb{S}\to[0,\infty]$*, defined by* (4.7)*, where*

$$\mathscr{G}^0\Big(\xi,\int_0^\cdot\dot h(s)\mathrm{d}s,\nu_T^\varphi\Big):=Y^u(\cdot;\xi),\qquad \xi\in\mathscr{C},\ u=(h,\varphi)\in S, \tag{4.10}$$

satisfies the skeleton equation

$$\mathrm{d}\{Y^u(t;\xi)-G(Y_t^u(\xi))\}=\Big\{b(t,Y_t^u(\xi))+\sigma(t,Y_t^u(\xi))\dot h(t) \tag{4.11}$$
$$+\int_{\mathbb{X}}\phi(t,Y_t^u(\xi),x)(\varphi(t,x)-1)\nu(\mathrm{d}x)\Big\}\mathrm{d}t$$

with the initial data $Y_0^u(\xi)=\xi\in\mathscr{C}$.

To highlight the initial data $X_0 = \xi \in \mathscr{C}$, we write the solution process of (4.8) as $X^\varepsilon(t;\xi)$ in lieu of $X^\varepsilon(t)$. By the Yamada–Watanabe theorem, there exists a measurable map $\mathscr{G}^\epsilon : \mathscr{C} \times \overline{\mathbb{V}} \mapsto \mathbb{S}$ such that

$$X^\varepsilon(\cdot;\xi) = \mathscr{G}^\varepsilon(\xi, \sqrt{\varepsilon}W, \varepsilon N^{\varepsilon^{-1}}).$$

Then, for any $\xi_\varepsilon \in \mathscr{C}$ and $u_\varepsilon = (h_\varepsilon, \varphi_\varepsilon) \in \overline{\mathscr{U}}_N$, by the Girsanov theorem [26, Lemma 2.3],

$$X^\varepsilon(\cdot;\xi_\varepsilon, u_\varepsilon) := \mathscr{G}^\varepsilon\Big(\xi_\varepsilon, \sqrt{\varepsilon}W + \int_0^\cdot \dot{h}_\varepsilon(s)\mathrm{d}s, \varepsilon N^{\varepsilon^{-1}\varphi_\varepsilon}\Big) \tag{4.12}$$

uniquely solves the following equation:

$$\begin{aligned} &\mathrm{d}\{X^\varepsilon(t;\xi_\varepsilon, u_\varepsilon) - G(X_t^\varepsilon(\xi_\varepsilon, u_\varepsilon))\} \\ &\quad = \{b(t, X_t^\varepsilon(\xi_\varepsilon, u_\varepsilon)) + \sigma(t, X_t^\varepsilon(\xi_\varepsilon, u_\varepsilon))\dot{h}_\varepsilon(t)\}\mathrm{d}t \\ &\quad\quad + \sqrt{\varepsilon}\sigma(t, X_t^\varepsilon(\xi_\varepsilon, u_\varepsilon))\mathrm{d}W(t) \\ &\quad\quad + \int_{\mathbb{X}} \phi(t, X_{t-}^\varepsilon(\xi_\varepsilon, u_\varepsilon), x)(\varepsilon N^{\varepsilon^{-1}\varphi_\varepsilon}(\mathrm{d}t, \mathrm{d}x) - \nu_T(\mathrm{d}t, \mathrm{d}x)), \quad t > 0 \end{aligned} \tag{4.13}$$

with the initial value $X_0^\varepsilon(\xi_\varepsilon, u_\varepsilon) = \xi_\varepsilon \in \mathscr{C}$. The argument of Theorem 4.7 is based on the following auxiliary lemmas. In what follows, we assume that K is a compact subset of $\mathscr{C} =: \mathbb{S}_0$.

Lemma 4.8 *Let* (H1)–(H3). *Then, there exists* $\varepsilon_0 \in (0, 1)$ *such that*

$$\overline{\mathbb{E}}\Big(\sup_{0\le t\le T} |X^\varepsilon(t;\xi, u_\varepsilon) - X^\varepsilon(t;\eta, u_\varepsilon)|^2\Big) \lesssim \|\xi - \eta\|_\infty^2, \quad \varepsilon \in (0, \varepsilon_0) \tag{4.14}$$

for any $\xi, \eta \in K$ *and* $u_\varepsilon = (h_\varepsilon, \varphi_\varepsilon) \in \overline{\mathscr{U}}_N$.

Proof For notational simplicity, let $Z^\varepsilon(t) = X^\varepsilon(t;\xi, u_\varepsilon) - X^\varepsilon(t;\eta, u_\varepsilon)$, and

$$M^\varepsilon(t) = X^\varepsilon(t;\xi, u_\varepsilon) - X^\varepsilon(t;\eta, u_\varepsilon) - \{G(X_t^\varepsilon(\xi, u_\varepsilon)) - G(X_t^\varepsilon(\eta, u_\varepsilon))\}.$$

According to the elementary inequality (1.21), one has

$$\sup_{0\le s\le t} |Z^\varepsilon(s)|^2 \lesssim \|\xi - \eta\|_\infty^2 + \sup_{0\le s\le t} |M^\varepsilon(s)|^2. \tag{4.15}$$

On the other hand, observe from (H1) that

$$|M^\varepsilon(t)| \lesssim \|Z_t^\varepsilon\|_\infty. \tag{4.16}$$

By the Itô formula, it follows that

$$
\begin{aligned}
&|M^{\varepsilon}(t)|^2 \\
&\leq |M^{\varepsilon}(0)|^2 + \int_0^t \{2|M^{\varepsilon}(s)|(|b(s, X_s^{\varepsilon}(\xi, u_\varepsilon)) - b(s, X_s^{\varepsilon}(\eta, u_\varepsilon)) \\
&+ (\sigma(s, X_s^{\varepsilon}(\xi, u_\varepsilon)) - \sigma(s, X_s^{\varepsilon}(\eta, u_\varepsilon)))\dot{h}_\varepsilon(s)|) \\
&+ \varepsilon \|\sigma(s, X_s^{\varepsilon}(\xi, u_\varepsilon)) - \sigma(s, X_s^{\varepsilon}(\eta, u_\varepsilon))\|_{HS}^2\} \mathrm{d}s \qquad (4.17) \\
&+ \varepsilon^2 \int_0^t \int_{\mathbb{X}} |\phi(s, X_{s-}^{\varepsilon}(\xi, u_\varepsilon), x) - \phi(s, X_{s-}^{\varepsilon}(\eta, u_\varepsilon), x)|^2 N^{\varepsilon^{-1}\varphi_\varepsilon}(\mathrm{d}s, \mathrm{d}x) \\
&+ 2\int_0^t \int_{\mathbb{X}} \{|M^{\varepsilon}(s)| \cdot |\phi(s, X_s^{\varepsilon}(\xi, u_\varepsilon), x) - \phi(s, X_s^{\varepsilon}(\eta, u_\varepsilon), x)| \\
&\times |1 - \varphi_\varepsilon(s, x)|\}\nu(\mathrm{d}x)\mathrm{d}s + M_{1,\varepsilon}(t) + M_{2,\varepsilon}(t),
\end{aligned}
$$

where

$$
M_{1,\varepsilon}(t) := 2\sqrt{\varepsilon} \sup_{0 \leq s \leq t} \left| \int_0^s \langle M^{\varepsilon}(z), (\sigma(z, X_z^{\varepsilon}(\xi, u_\varepsilon)) - \sigma(z, X_z^{\varepsilon}(\eta, u_\varepsilon)))\mathrm{d}W(z)\rangle \right|,
$$

and

$$
\begin{aligned}
M_{2,\varepsilon}(t) :&= 2\varepsilon \sup_{0 \leq s \leq t} \left| \int_0^s \int_{\mathbb{X}} \langle M^{\varepsilon}(z-), \phi(z, X_{z-}^{\varepsilon}(\xi, u_\varepsilon), x) - \phi(z, X_{z-}^{\varepsilon}(\eta, u_\varepsilon), x)\rangle \right. \\
&\left. \times (N^{\varepsilon^{-1}\varphi_\varepsilon}(\mathrm{d}z, \mathrm{d}x) - \varepsilon^{-1}\varphi_\varepsilon(z, x)\nu(\mathrm{d}x)\mathrm{d}z) \right|.
\end{aligned}
$$

Recall from [24, Remark 3.6] that

$$
\sup_{\varphi \in S_{2,N}} \int_{\mathbb{X}_T} \Gamma_1^2(t, x)|1 - \varphi(t, x)|\nu_T(\mathrm{d}t, \mathrm{d}x) < \infty, \tag{4.18}
$$

and

$$
\sup_{\varphi \in S_{2,N}} \int_{\mathbb{X}_T} \Gamma_1(t, x)|1 - \varphi(t, x)|\nu_T(\mathrm{d}t, \mathrm{d}x) < \infty. \tag{4.19}
$$

In view of (H2), (4.15)–(4.17), and (4.19), we obtain that

$$
\begin{aligned}
&\sup_{0\le s\le t}|Z^\varepsilon(s)|^2\\
&\lesssim\Big(1+\int_0^t\int_{\mathbb{X}}\Gamma_1(s,x)|1-\varphi_\varepsilon(s,x)|\nu_T(\mathrm{d}s,\mathrm{d}x)\Big)\|\xi-\eta\|_\infty^2+M_{1,\varepsilon}(t)+M_{2,\varepsilon}(t)\\
&\quad+\varepsilon^2\int_0^t\int_{\mathbb{X}}|\phi(s,X^\varepsilon_{s-}(\xi,u_\varepsilon),x)-\phi(s,X^\varepsilon_{s-}(\eta,u_\varepsilon),x)|^2N^{\varepsilon^{-1}\varphi_\varepsilon}(\mathrm{d}s,\mathrm{d}x)\\
&\quad+\int_0^t\Big\{1+|\dot h_\varepsilon(s)|^2+\int_{\mathbb{X}}\Gamma_1(s,x)|1-\varphi_\varepsilon(s,x)|\nu(\mathrm{d}x)\Big\}\sup_{0\le r\le s}|Z^\varepsilon(r)|^2\mathrm{d}s\\
&\lesssim\|\xi-\eta\|_\infty^2+M_{1,\varepsilon}(t)+M_{2,\varepsilon}(t)\\
&\quad+\varepsilon^2\int_0^t\int_{\mathbb{X}}|\phi(s,X^\varepsilon_{s-}(\xi,u_\varepsilon),x)-\phi(s,X^\varepsilon_{s-}(\eta,u_\varepsilon),x)|^2N^{\varepsilon^{-1}\varphi_\varepsilon}(\mathrm{d}s,\mathrm{d}x)\\
&\quad+\int_0^t\sup_{0\le r\le s}|Z^\varepsilon(r)|^2\mathrm{d}s.
\end{aligned}
$$

This, combined with the Gronwall inequality, gives that

$$
\begin{aligned}
&\overline{\mathbb{E}}\Big(\sup_{0\le s\le t}|Z^\varepsilon(s)|^2\Big)\\
&\lesssim\|\xi-\eta\|_\infty^2+\overline{\mathbb{E}}M_{1,\varepsilon}(t)+\overline{\mathbb{E}}M_{2,\varepsilon}(t) \qquad (4.20)\\
&\quad+\varepsilon^2\overline{\mathbb{E}}\int_0^t\int_{\mathbb{X}}|\phi(s,X^\varepsilon_{s-}(\xi,u_\varepsilon),x)-\phi(s,X^\varepsilon_{s-}(\eta,u_\varepsilon),x)|^2N^{\varepsilon^{-1}\varphi_\varepsilon}(\mathrm{d}s,\mathrm{d}x)\\
&=:\|\xi-\eta\|_\infty^2+\Theta(t)
\end{aligned}
$$

It is easily seen from (H2) that

$$
\int_{\mathbb{X}_T}\Gamma_1^2(t,x)\nu_T(\mathrm{d}t,\mathrm{d}x)\lesssim T. \tag{4.21}
$$

Furthermore, by the B–D–G inequality, the Jensen inequality, in addition to (4.16) and (H2), we deduce that

$$
\begin{aligned}
\Theta(t)&\lesssim\sqrt{\varepsilon}\overline{\mathbb{E}}\Big(\int_0^t\|Z_s^\varepsilon\|_\infty^4\mathrm{d}s\Big)^{1/2}+\varepsilon\overline{\mathbb{E}}\int_0^t\int_{\mathbb{X}}\|Z_s^\varepsilon\|_\infty^2\Gamma_1^2(s,x)\varphi_\varepsilon(s,x)\nu(\mathrm{d}x)\mathrm{d}s\\
&\quad+\varepsilon\overline{\mathbb{E}}\Big(\int_0^t\int_{\mathbb{X}}\|Z_s^\varepsilon\|_\infty^4\Gamma_1^2(s,x)N^{\varepsilon^{-1}\varphi_\varepsilon}(\mathrm{d}s,\mathrm{d}x)\Big)^{1/2}\\
&\lesssim\sqrt{\varepsilon}\overline{\mathbb{E}}\Big(\int_0^t\|Z_s^\varepsilon\|_\infty^4\mathrm{d}s\Big)^{1/2}+\varepsilon\overline{\mathbb{E}}\int_0^t\int_{\mathbb{X}}\|Z_s^\varepsilon\|_\infty^2\Gamma_1^2(s,x)\varphi_\varepsilon(s,x)\nu(\mathrm{d}x) \qquad (4.22)\\
&\quad+\sqrt{\varepsilon}\Big(\overline{\mathbb{E}}\int_0^t\int_{\mathbb{X}}\|Z_s^\varepsilon\|_\infty^4\Gamma_1^2(s,x)\varphi_\varepsilon(s,x)\nu(\mathrm{d}x)\mathrm{d}s\Big)^{1/2}\\
&\lesssim\|\xi-\eta\|_\infty^2+\sqrt{\varepsilon}\,\overline{\mathbb{E}}\Big(\sup_{0\le s\le t}|Z^\varepsilon(s)|^2\Big),
\end{aligned}
$$

where in the last step we have used (4.18) and (4.21). Substituting (4.22) into (4.20) leads to

$$\begin{aligned}\bar{\mathbb{E}}\Big(\sup_{0\le s\le t}|Z^\varepsilon(s)|^2\Big) &\lesssim \|\xi-\eta\|_\infty^2+\sqrt{\varepsilon}\,\bar{\mathbb{E}}\Big(\sup_{0\le s\le t}|Z^\varepsilon(s)|^2\Big)\\ &\lesssim \|\xi-\eta\|_\infty^2+\sqrt{\varepsilon_0}\,\bar{\mathbb{E}}\Big(\sup_{0\le s\le t}|Z^\varepsilon(s)|^2\Big).\end{aligned}$$

Consequently, the desired assertion (4.14) follows immediately.

Set $M_\varepsilon(t) := X^\varepsilon(t;\xi,u_\varepsilon) - G(X_t^\varepsilon(\xi,u_\varepsilon))$. Note from (1.21) and (H1) that

$$\sup_{0\le s\le t}|X^\varepsilon(s;\xi,u_\varepsilon)|^2 \lesssim 1+\|\xi\|_\infty^2+\sup_{0\le s\le t}|M_\varepsilon(s)|^2,$$

and

$$|M_\varepsilon(t)| \lesssim 1+\|X_t^\varepsilon(\xi,u_\varepsilon)\|_\infty.$$

Then, carrying out a similar argument to derive Lemma 4.8 gives the following estimate.

Lemma 4.9 *Under the assumptions of Lemma* 4.8,

$$\bar{\mathbb{E}}\Big(\sup_{0\le t\le T}|X^\epsilon(t;\xi,u_\varepsilon)|^2\Big)<\infty. \tag{4.23}$$

Lemma 4.10 *Let* (H1)–(H3) *hold and assume further that* $u_n=(h_n,\varphi_n)\in S_N$, $u=(h,\varphi)\in S_N$ *and* $\xi_n,\xi\in K$ *such that* $u_n\to u$ *and* $\xi_n\to\xi$ *as* $n\to\infty$. *Then,*

$$\mathscr{G}^0\Big(\xi_n,\int_0^\cdot \dot{h}_n(s)\mathrm{d}s,\nu_T^{\varphi_n}\Big)\to\mathscr{G}^0\Big(\xi,\int_0^\cdot \dot{h}(s)\mathrm{d}s,\nu_T^{\varphi}\Big)\quad in\ C([0,T];\mathbb{R}^n),$$

where $\mathscr{G}^0$ *is defined in* (4.10).

Proof To complete the proof, by the triangle inequality, it suffices to show that

(I) $\lim_{n\to\infty}\sup_{0\le t\le T}|Y^{u_n}(t;\xi_n)-Y^{u_n}(t;\xi)|=0$;
(II) $\lim_{n\to\infty}\sup_{0\le t\le T}|Y^{u_n}(t;\xi)-Y^{u}(t;\xi)|=0$.

Following an argument as that of Lemma 4.8, we infer that (I) holds. According to the Arzelà–Ascoli theorem, to verify the claim (II), we shall prove that

(a) $(Y^{u_n}(\cdot,\xi))_{n\in\mathbb{N}}$ is uniformly bounded, i.e., $\sup_n\sup_{t\in[0,T]}|Y^{u_n}(t;\xi)|<\infty$;
(b) $(Y^{u_n}(\cdot,\xi))_{n\in\mathbb{N}}$ is uniformly equicontinuous, i.e.,

$$\lim_{\delta\to 0}\sup_n|Y^{u_n}(t+\delta;\xi)-Y^{u_n}(t;\xi)|=0.$$

The uniform boundedness (i.e., (a)) follows from a similar argument of Lemma 4.8. Next, we claim the uniform equicontinuity (i.e., (b)). In what follows, without loss of generality, let $\delta \in (0, \tau)$. Set

$$M^n(t) := Y^{u_n}(t;\xi) - G(Y_t^{u_n}(\xi)).$$

Note from (H1) and (1.21) that

$$\begin{aligned}&\sup_{0\le t\le T}|Y^{u_n}(t+\delta) - Y^{u_n}(t)|^2 \\ &\lesssim \sup_{-\tau\le t\le -\delta}|\xi(t+\delta)-\xi(t)|^2 + \sup_{-\delta\le t\le 0}|\xi(0)-\xi(t)|^2 \\ &\quad + \sup_{-\delta\le t\le 0}|Y^{u_n}(t+\delta;\xi)-\xi(0)|^2 + \sup_{0\le t\le T}|M^n(t+\delta)-M^n(t)|^2.\end{aligned} \tag{4.24}$$

By the chain rule, it follows from (a) that

$$|M^n(t+\delta)-M^n(t)|^2 \lesssim \int_t^{t+\delta}(1+|\dot{h}_n(s)|^2)\mathrm{d}s + \int_t^{t+\delta}\int_{\mathbb{X}}\Gamma_0(s,x)|1-\varphi_n(s,x)|\nu(\mathrm{d}x)\mathrm{d}s.$$

In view of (4.9) and (H3), we get from [24, (3.5), Lemma 3.4] that

$$\lim_{\delta\to 0}\sup_{\varphi\in S_{2,N}}\sup_{|s-t|\le\delta}\int_{[s,t]\times\mathbb{X}}\Gamma_0(r,x)|\varphi(r,x)-1|\nu(\mathrm{d}x)\mathrm{d}r = 0, 0\le s\le t. \tag{4.25}$$

This, together with $h_n \in S_{1,N}$, implies that

$$\lim_{\delta\to 0}\sup_{0\le t\le T}|M^n(t+\delta)-M^n(t)|^2 = 0. \tag{4.26}$$

Analogically,

$$\lim_{\delta\to 0}\sup_{-\delta\le t\le 0}|Y^{u_n}(t+\delta;\xi)-\xi(0)|^2 = 0. \tag{4.27}$$

As a result, (b) follows from (4.24)–(4.27) and $\xi\in\mathscr{C}$.

Since $(Y^{u_n}(\cdot;\xi))_{n\in\mathbb{N}}$ is pre-compact in $C([0,T];\mathbb{R}^n)$, every subsequence, which is still denoted by $(Y^{u_n}(\cdot;\xi))_{n\in\mathbb{N}}$, has a convergent subsequence. Let $\tilde{Y}(\cdot,\xi)$ be the limit point of $(Y^{u_n}(\cdot;\xi))_{n\in\mathbb{N}}$. Observe that

$$\begin{aligned}&\int_{\mathbb{X}_T}\{\phi(t,Y_t^{u_n}(\xi),x)(\varphi_n(t,x)-1)-\phi(t,\tilde{Y}_t(\xi),x)(\varphi(t,x)-1)\}\nu_T(\mathrm{d}t,\mathrm{d}x)\\ &\le \int_{\mathbb{X}_T}\|Y_t^{u_n}(\xi)-\tilde{Y}_t(\xi)\|_\infty\Gamma_1(t,x)|\varphi_n(t,x)-1|\nu_T(\mathrm{d}t,\mathrm{d}x)\\ &\quad+\int_{\mathbb{X}_T}(1+\|\tilde{Y}_t(\xi)\|_\infty)\Gamma_0(t,x)|\varphi_n(t,x)-\varphi(t,x)|\nu_T(\mathrm{d}t,\mathrm{d}x),\end{aligned} \tag{4.28}$$

and, from [24, Lemma 3.11], that

$$\int_{\mathbb{X}_T} \Gamma_0(t,x)(\varphi_n(t,x)-1)\nu_T(\mathrm{d}t,\mathrm{d}x) \to \int_{\mathbb{X}_T} \Gamma_0(t,x)(\varphi(t,x)-1)\nu_T(\mathrm{d}t,\mathrm{d}x) \quad (4.29)$$

as $n \to \infty$. By the dominated convergence theorem, in addition to (4.19), (4.28), and (4.29), we conclude that $\tilde{Y}(\cdot;\xi) = Y^u(\cdot;\xi)$ from the uniqueness. Consequently, (II) follows.

The following tightness criteria plays an important role in establishing tightness of a sequence with càdlàg paths.

Lemma 4.11 ([1, Theorem 1]) *Let (Λ^ε) be a sequence of random elements of $D([0,T];\mathbb{R}^n)$ and $\{\tau_\varepsilon, \delta_\varepsilon\}$ such that*

(i) *For each $\varepsilon > 0$, τ_ε is a stopping time with respect to the natural σ-algebra, and takes only finitely many values;*
(ii) *For each $\varepsilon > 0$, δ_ε is a constant, $0 \le \delta_\varepsilon \le T$, and $\delta_\varepsilon \to 0$ as $\varepsilon \to 0$.*

Then, (Λ^ε) is tight in $D([0,T];\mathbb{R}^n)$ if

(I) *$\Lambda^\varepsilon(t)$ is tight on the line for each $t \in [0,T]$, and*
(II) *$\Lambda^\varepsilon(\tau_\varepsilon + \delta_\varepsilon) - \Lambda^\varepsilon(\tau_\varepsilon) \to 0$ in probability as $\varepsilon \to 0$.*

Lemma 4.12 *Assume that $u_\varepsilon = (h_\varepsilon, \varphi_\varepsilon) \in \overline{\mathscr{U}}_N$, $u = (h,\varphi) \in \overline{\mathscr{U}}_N$ and $\xi_\varepsilon, \xi \in K$ such that $u_\varepsilon \Rightarrow u$, under $\overline{\mathbb{P}}$, and $\xi_\varepsilon \to \xi$ as $\varepsilon \to 0$. Then*

$$\mathscr{G}^\varepsilon(\xi_\varepsilon, \sqrt{\varepsilon}W + \int_0^\cdot \dot{h}_\varepsilon(s)\mathrm{d}s, \varepsilon N^{\varepsilon^{-1}\varphi_\varepsilon}) \Rightarrow \mathscr{G}^0\Big(\xi, \int_0^\cdot \dot{h}(s)\mathrm{d}s, \nu_T^\varphi\Big),$$

under $\overline{\mathbb{P}}$, as $\varepsilon \to 0$, where $\mathscr{G}^\varepsilon$ satisfying (4.12).

Proof By Lemma 4.8, we need only show that

$$\begin{aligned} X^\varepsilon(\cdot;\xi,u_\varepsilon) &= \mathscr{G}^\varepsilon\Big(\xi, \sqrt{\varepsilon}W + \int_0^\cdot \dot{h}_\varepsilon(s)\mathrm{d}s, \varepsilon N^{\varepsilon^{-1}\varphi_\varepsilon}\Big) \qquad (4.30)\\ &\Rightarrow \mathscr{G}^0\Big(\xi, \int_0^\cdot \dot{h}(s)\mathrm{d}s, \nu_T^\varphi\Big) \end{aligned}$$

under $\overline{\mathbb{P}}$ as $\varepsilon \to 0$. We first show that $(X^\varepsilon(\cdot;\xi,u_\varepsilon))_{\varepsilon>0}$ is tight in $D([0,T];\mathbb{R}^n)$. To this end, it is sufficient to verify that (I) and (II) in Lemma 4.11 hold, respectively, for $(X^\varepsilon(\cdot;\xi,u_\varepsilon))_{\varepsilon>0}$. By (4.23) and the Chebyshev inequality, for any $R > 0$,

$$\overline{\mathbb{P}}\Big(\sup_{0\le t\le T} |X^\varepsilon(t;\xi,u_\varepsilon)| > R\Big) \lesssim 1/R^2.$$

We conclude that $(X^\varepsilon(t;\xi,u_\varepsilon))_{\varepsilon>0}$ is tight on the line for each $t\in[0,T]$. Hence, (I) holds for $(X^\varepsilon(t;\xi,u_\varepsilon))_{\varepsilon>0}$. In what follows, without loss of generality, we may assume $\tau_\varepsilon\geq\tau$. For any $\{\tau_\varepsilon,\delta_\varepsilon\}$ such that (i) and (ii) in Lemma 4.11, observe from (1.21) and (H1) that

$$\begin{aligned}&\overline{\mathbb{E}}\|X^{\varepsilon,u_\varepsilon}_{\tau_\varepsilon+\delta_\varepsilon}-X^{\varepsilon,u_\varepsilon}_{\tau_\varepsilon}\|_\infty\\&\leq\frac{M_{\tau_\varepsilon,\delta_\varepsilon}}{1-\kappa}+\frac{\kappa}{1-\kappa}\overline{\mathbb{E}}\|X^\varepsilon_{\tau_\varepsilon-\tau+\delta_\varepsilon}(\xi,u_\varepsilon)-X^\varepsilon_{\tau_\varepsilon-\tau}(\xi,u_\varepsilon)\|_\infty,\end{aligned}\tag{4.31}$$

where $M_{\tau_\varepsilon,\delta_\varepsilon}:=\overline{\mathbb{E}}\Big(\sup_{\tau_\varepsilon-\tau\leq r\leq\tau_\varepsilon}|M_\varepsilon(r+\delta_\varepsilon;\xi,u_\varepsilon)-M_\varepsilon(r;\xi,u_\varepsilon)|\Big)$. Next, it is readily seen that

$$\begin{aligned}&M_{\tau_\varepsilon,\delta_\varepsilon}\\&\leq\overline{\mathbb{E}}\Big(\sup_{\tau_\varepsilon-\tau\leq t\leq\tau_\varepsilon}\int_t^{t+\delta_\varepsilon}|b(s,X^\varepsilon_s(\xi,u_\varepsilon))+\sigma(s,X^\varepsilon_s(\xi,u_\varepsilon))\dot h_\varepsilon(s)|\mathrm{d}s\Big)\\&\quad+\overline{\mathbb{E}}\Big(\sup_{\tau_\varepsilon-\tau\leq t\leq\tau_\varepsilon}\int_t^{t+\delta_\varepsilon}\int_{\mathbb{X}}|\phi(s,X^\varepsilon_s(\xi,u_\varepsilon),x)|\cdot|\varphi_\varepsilon(s,x)-1|\nu(\mathrm{d}x)\mathrm{d}s\Big)\\&\quad+\sqrt{\varepsilon}\,\overline{\mathbb{E}}\Big(\sup_{\tau_\varepsilon-\tau\leq t\leq\tau_\varepsilon}\Big|\int_t^{t+\delta_\varepsilon}\sigma(s,X^\varepsilon_s(\xi,u_\varepsilon))\mathrm{d}W(s)\Big|\Big)\\&\quad+\varepsilon\overline{\mathbb{E}}\Big(\sup_{\tau_\varepsilon-\tau\leq t\leq\tau_\varepsilon}\Big|\int_t^{t+\delta_\varepsilon}\int_{\mathbb{X}}\phi(s,X^\varepsilon_{s-}(\xi,u_\varepsilon),x)\tilde N^{\varepsilon^{-1}\varphi_\varepsilon}(\mathrm{d}s,\mathrm{d}x)\Big|\Big)\\&=:\Xi_1+\Xi_2+\Xi_3+\Xi_4.\end{aligned}\tag{4.32}$$

By the Hölder inequality and (4.9), we obtain from (4.23) that

$$\begin{aligned}\Xi_1+\Xi_2\lesssim\sqrt{\delta_\varepsilon}&\Big\{1+\overline{\mathbb{E}}\Big(\sup_{0\leq t\leq T}\|X^\varepsilon_t(\xi,u_\varepsilon)\|_\infty\Big)\Big\}\\&+\overline{\mathbb{E}}\Big\{\Big(1+\sup_{0\leq t\leq T}\|X^\varepsilon_t(\xi,u_\varepsilon)\|_\infty\Big)\\&\times\sup_{\tau_\varepsilon-\tau\leq t\leq\tau_\varepsilon}\int_t^{t+\delta_\varepsilon}\int_{\mathbb{X}}\Gamma_0(s,x)|\varphi_\varepsilon(s,x)-1|\nu(\mathrm{d}x)\mathrm{d}s\Big\}.\end{aligned}\tag{4.33}$$

Observe from [24, (3.3), Lemma 3.4] that

$$\sup_{\varphi_\varepsilon\in S_{2,N}}\int_{\mathbb{X}_T}\Gamma_0^2(t,x)(1+\varphi_\varepsilon(t,x))\nu_T(\mathrm{d}t,\mathrm{d}x)<\infty.\tag{4.34}$$

Furthermore, according to the B–D–G inequality and the Jensen inequality together with (4.9), we deduce that

$$\Xi_3 + \Xi_4 \lesssim \sqrt{\varepsilon}\Big\{\overline{\mathbb{E}}\Big(1 + \sup_{0\le t\le T} \|X_t^\varepsilon(\xi, u_\varepsilon)\|_\infty^2\Big) \times \int_0^T \Big(1 + \int_{\mathbb{X}} \Gamma_0^2(t,x)\varphi_\varepsilon(t,x)\nu(\mathrm{d}x)\Big)\mathrm{d}t\Big\}^{1/2}. \quad (4.35)$$

Inserting (4.32)–(4.35) into (4.31), and taking (4.23) and (4.34) into account yields

$$\begin{aligned}&\overline{\mathbb{E}}\|X^\varepsilon_{\tau_\varepsilon+\delta_\varepsilon}(\xi,u_\varepsilon) - X^\varepsilon_{\tau_\varepsilon}(\xi,u_\varepsilon)\|_\infty\\ &\lesssim \sqrt{\varepsilon} + \sqrt{\delta_\varepsilon} + \frac{\kappa}{1-\kappa}\overline{\mathbb{E}}\|X^\varepsilon_{\tau_\varepsilon-\tau+\delta_\varepsilon}(\xi,u_\varepsilon) - X^\varepsilon_{\tau_\varepsilon-\tau}(\xi,u_\varepsilon)\|_\infty\\ &+ \overline{\mathbb{E}}\Big\{\Big(1 + \sup_{0\le t\le T}\|X_t^\varepsilon(\xi,u_\varepsilon)\|_\infty\Big)\sup_{\tau_\varepsilon-\tau\le t\le\tau_\varepsilon}\int_t^{t+\delta_\varepsilon}\int_{\mathbb{X}}\Gamma_0(s,x)|\varphi_\varepsilon(s,x) - 1|\nu_T(\mathrm{d}s,\mathrm{d}x)\Big\}.\end{aligned}$$

Then, for any $\eta > 0$, by (4.25), there is an $\varepsilon_0 = \varepsilon_0(\eta) > 0$ such that, for $\varepsilon \in (0, \varepsilon_0)$,

$$\begin{aligned}&\overline{\mathbb{E}}\|X^\varepsilon_{\tau_\varepsilon+\delta_\varepsilon}(\xi,u_\varepsilon) - X^\varepsilon_{\tau_\varepsilon}(\xi,u_\varepsilon)\|_\infty\\ &\lesssim \sqrt{\varepsilon} + \sqrt{\delta_\varepsilon} + \eta + \frac{\kappa}{1-\kappa}\overline{\mathbb{E}}\|X^\varepsilon_{\tau_\varepsilon-\tau+\delta_\varepsilon}(\xi,u_\varepsilon) - X^\varepsilon_{\tau_\varepsilon-\tau}(\xi,u_\varepsilon)\|_\infty.\end{aligned} \quad (4.36)$$

Then, by an induction argument, in addition to the continuity of initial data and the arbitrariness of η, we conclude from the Chebyshev inequality that (II) (in Lemma 4.11) holds for $(X^\varepsilon(\cdot,\xi,u_\varepsilon))$. So, by Lemma 4.11, $(X^\varepsilon(\cdot;\xi,u_\varepsilon))$ is tight in $D([0,T];\mathbb{R}^n)$.

In what follows, it suffices to show that $\mathscr{G}^0\Big(\xi, \int_0^\cdot \dot{h}(s)\mathrm{d}s, \nu_T^\varphi\Big)$ is the unique limit point of $\mathscr{G}^\varepsilon(\xi, \sqrt{\varepsilon}W + \int_0^\cdot \dot{h}_\varepsilon(s)\mathrm{d}s, \varepsilon N^{\varepsilon^{-1}\varphi_\varepsilon})$. Let

$$M^\varepsilon(t) = \sqrt{\varepsilon}\int_0^t \sigma(s, X_s^\varepsilon(\xi,u_\varepsilon))\mathrm{d}W(s), \quad \text{and}$$
$$\overline{M}^\varepsilon(t) = \varepsilon\int_0^t\int_{\mathbb{X}} \phi(s, X^\varepsilon_{s-}(\xi,u_\varepsilon), x)\tilde{N}^{\varepsilon^{-1}\varphi_\varepsilon}(\mathrm{d}s,\mathrm{d}x).$$

We can choose a subsequence of $(X^\varepsilon(\cdot;\xi,u_\varepsilon), u_\varepsilon, M^\varepsilon, \overline{M}^\varepsilon)$ convergent to $(Y, u, 0, 0)$ in distribution as $\varepsilon \to 0$ since $(X^\varepsilon(\cdot,\xi,u_\varepsilon))$ is tight in $D([0,T];\mathbb{R}^n)$. Without loss of generality, by the Skorokhod representation theorem, we may assume

$$(X^\varepsilon(\cdot,\xi,u_\varepsilon), u_\varepsilon, M^\varepsilon, \overline{M}^\varepsilon) \to (Y, u, 0, 0), \quad \overline{\mathbb{P}} - \text{a. s.}$$

Letting $\varepsilon \to 0$ on both sides of (4.13), Y satisfies (4.11). Then, the desired assertion follows from the uniqueness.

Proof of Theorem 4.7

Proof With Lemma 4.5 in hand, the argument of Theorem 4.7 follows from Lemmas 4.10 and 4.12 immediately.

4.4 Uniform Large Deviations under Polynomial Growth Conditions

In this section, we are concerned with the uniform LDPs of the following SDDE of neutral type

$$\begin{aligned} &\mathrm{d}\{X^\varepsilon(t) - G(X^\varepsilon(t-\tau))\} \\ &= b(t, X^\varepsilon(t), X^\varepsilon(t-\tau))\mathrm{d}t \\ &+ \sqrt{\varepsilon}\sigma(t, X^\varepsilon(t), X^\varepsilon(t-\tau))\mathrm{d}W(t) \\ &+ \varepsilon\int_{\mathbb{X}} \phi(t, X^\varepsilon(t-), X^\varepsilon((t-\tau)-), x)\tilde{N}^{\varepsilon^{-1}}(\mathrm{d}t, \mathrm{d}x) \end{aligned} \tag{4.37}$$

with the initial datum $X_0^\varepsilon = \xi \in \mathscr{C}$, where $G : \mathbb{R}^n \mapsto \mathbb{R}^n$, $b : [0,\infty)\times\mathbb{R}^n\times\mathbb{R}^n \mapsto \mathbb{R}^n$, $\sigma : [0,\infty)\times\mathbb{R}^n\times\mathbb{R}^n \mapsto \mathbb{R}^n \otimes \mathbb{R}^m$, and $\phi : [0,\infty)\times\mathbb{R}^n\times\mathbb{R}^n\times\mathbb{X} \mapsto \mathbb{R}^n$. We assume that the following assumptions on the coefficients of (4.37) hold.

(A1) There exist $\lambda_1 > 0$ and $q \geq 1$ such that

$$|G(x) - G(y)|^2 \leq \lambda_1(1 + |x|^q + |y|^q)|x-y|^2, \qquad x, y \in \mathbb{R}^n.$$

(A2) There exists a $\lambda_2 > 0$ such that for any $x_1, x_2, y_1, y_2 \in \mathbb{R}^n$ and $x \in \mathbb{X}$,

$$\begin{aligned} &|b(t, x_1, y_1) - b(t, x_2, y_2)|^2 + \|\sigma(t, x_1, y_1) - \sigma(t, x_2, y_2)\|_{HS}^2 \\ &\qquad \leq \lambda_2(|x_1 - x_2|^2 + (1 + |y_1|^q + |y_2|^q)|y_1 - y_2|^2), \end{aligned}$$

and

$$\begin{aligned} |\phi(t, x_1, y_1, x) - \phi(t, x_2, y_2, x)|^2 &\leq \lambda_2(|x_1 - x_2|^2 \\ &\quad + (1 + |y_1|^q + |y_2|^q)|y_1 - y_2|^2)\|x\|_{\mathbb{X}}^2. \end{aligned}$$

(A3) Define $\tilde{\rho}_T = \sup_{0\leq t\leq T}\{|b(t,0,0)|^2 + \|\sigma(t,0,0)\|_{HS}^2\}$. Assume that $\tilde{\rho}_T < \infty$ and that there is a $\lambda_3 > 0$ such that

$$\sup_{0\leq t\leq T} |\phi(t, 0, 0, x)|^2 \leq \lambda_3\|x\|_{\mathbb{X}}^2, \qquad x \in \mathbb{X}.$$

(A4) There exists a $\delta_1 \in (0, \infty)$ such that for any $p \geq 2$ and $E \in \mathscr{B}(\mathbb{X})$ with $\nu(E) < \infty$,

$$\int_E \exp(\delta_1 \|x\|_{\mathbb{X}}^p)\nu(\mathrm{d}x) < \infty.$$

Remark 4.13 Consider a scalar SDDE of neutral type

$$\begin{aligned} \mathrm{d}\{X^\varepsilon(t) - 2(X^\varepsilon(t-1))^2\} &= \{\sin(X^\varepsilon(t)) + X^\varepsilon(t-1)\}\mathrm{d}t \\ &\quad + \varepsilon \int_{\mathbb{X}} (X^\varepsilon((t-1)-))^3 N^{\varepsilon^{-1}}(\mathrm{d}t, \mathrm{d}x) \end{aligned} \tag{4.38}$$

with the initial data $X_0^\varepsilon = \xi \in \mathscr{C}$. For (4.38), it is easy to see that (H1) and (H2) no longer hold, but (A1) and (A2) do.

Remark 4.14 By (A1)–(A3), a straightforward calculation gives that for any $x, y \in \mathbb{R}^n$ and $z \in \mathbb{X}$,

$$|G(x)|^2 \lesssim 1 + |x|^{q+2}, \quad |b(t, x, y)|^2 + \|\sigma(t, x, y)\|_{HS}^2 \lesssim 1 + |x|^2 + |y|^{q+2}$$

and

$$|\phi(t, x, y, z)|^2 \lesssim (1 + |x|^2 + |y|^{q+2})\|z\|_{\mathbb{X}}^2.$$

Remark 4.15 For $t \in [0, \tau]$, (4.37) reduces to stochastic integral equation

$$\begin{aligned} X^\varepsilon(t) &= \xi(0) + G(\xi(t-\tau)) - G(\xi(-\tau)) + \int_0^t b(s, X^\varepsilon(s), \xi(s-\tau))\mathrm{d}s \\ &\quad + \sqrt{\varepsilon}\int_0^t \sigma(s, X^\varepsilon(s), \xi(s-\tau))\mathrm{d}W(s) \\ &\quad + \varepsilon \int_0^t \int_{\mathbb{X}} \phi(s, X^\varepsilon(s-), \xi(s-\tau), x)\tilde{N}^{\varepsilon^{-1}}(\mathrm{d}s, \mathrm{d}x). \end{aligned}$$

Let $\tilde{X}^\varepsilon(t) = X^\varepsilon(t) - G(\xi(t-\tau))$. Then, the above integral equation can be reformulated as

$$\begin{aligned} \tilde{X}^\varepsilon(t) &= \tilde{X}(0) + \int_0^t b(s, \tilde{X}^\varepsilon(s) + G(\xi(s-\tau)), \xi(s-\tau))\mathrm{d}s \\ &\quad + \sqrt{\varepsilon}\int_0^t \sigma(s, \tilde{X}^\varepsilon(s) + G(\xi(s-\tau)), \xi(s-\tau))\mathrm{d}W(s) \\ &\quad + \varepsilon \int_0^t \int_{\mathbb{X}} \phi(s, \tilde{X}^\varepsilon(s-) + G(\xi(s-\tau)), \xi(s-\tau), x)\tilde{N}^{\varepsilon^{-1}}(\mathrm{d}s, \mathrm{d}x). \end{aligned} \tag{4.39}$$

This is an SDE without delay and neutral term. Moreover, (A2)–(A4) guarantee the existence and uniqueness of (4.39). Repeating this argument, we conclude that (4.37) admits a unique global solution on $[0, T]$. Also, see Theorem A.7 for more details.

Our main result in this section is presented next.

Theorem 4.16 *Under* (A1)–(A4), $(X^\varepsilon(t;\xi))_{\varepsilon>0,t\in[0,T]}$ *satisfying* (4.37) *admits a uniform LDP in* $\mathbb{S} := D([0,T];\mathbb{R}^n)$ *with the rate function* $I_\xi : \mathbb{S} \to [0,\infty]$ *defined by* (4.7)*, where, for* $\xi \in \mathscr{C}$ *and* $u = (h,\varphi) \in S$,

$$\mathscr{G}^0\Big(\xi, \int_0^\cdot \dot{h}(s)\mathrm{d}s, \nu_T^\varphi\Big) := Y^u(\cdot;\xi) \tag{4.40}$$

and

$$\begin{aligned}
&\mathrm{d}\{Y^u(t;\xi) - G(Y^u(t-\tau;\xi))\}\\
&= \Big\{b(t, Y^u(t;\xi), Y^u(t-\tau;\xi)) + \sigma(t, Y^u(t;\xi), Y^u(t-\tau;\xi))\dot{h}(t)\\
&\quad + \int_{\mathbb{X}} \phi(t, Y^u(t;\xi), Y^u(t-\tau;\xi), x)(\varphi(t,x)-1)\nu(\mathrm{d}x)\Big\}\mathrm{d}t, \qquad t>0
\end{aligned} \tag{4.41}$$

with the initial data $X_0^u(\xi) = \xi \in \mathscr{C}$.

We follow the line of argument as that of the proof of Theorem 4.16. By the Yamada–Watanabe theorem, there exists a measurable map $\mathscr{G}^\epsilon : \mathscr{C} \times \overline{\mathbb{V}} \mapsto \mathbb{S}$ such that

$$X^\varepsilon(\cdot;\xi) = \mathscr{G}^\varepsilon(\xi, \sqrt{\varepsilon}W, \varepsilon N^{\varepsilon^{-1}}).$$

Then, for $\xi_\varepsilon \in \mathscr{C}$ and $u_\varepsilon = (h_\varepsilon, \varphi_\varepsilon) \in \overline{\mathscr{U}}_N$, by the Girsanov theorem (see, e.g., [126, Lemma 2.3]), we obtain that

$$X^\varepsilon(\cdot;\xi_\varepsilon,u_\varepsilon) := \mathscr{G}^\varepsilon\Big(\xi_\varepsilon, \sqrt{\varepsilon}W + \int_0^\cdot \dot{h}_\varepsilon(s)\mathrm{d}s, \varepsilon N^{\varepsilon^{-1}\varphi_\varepsilon}\Big) \tag{4.42}$$

uniquely solves

$$\begin{aligned}
&\mathrm{d}\{X^\varepsilon(t;\xi_\varepsilon,u_\varepsilon) - G(X^\varepsilon(t-\tau;\xi_\varepsilon,u_\varepsilon))\}\\
&= \{b(t, X^\varepsilon(t;\xi_\varepsilon,u_\varepsilon), X^\varepsilon(t-\tau;\xi_\varepsilon,u_\varepsilon))\\
&\quad + \sigma(t, X^\varepsilon(t;\xi_\varepsilon,u_\varepsilon), X^\varepsilon(t-\tau;\xi_\varepsilon,u_\varepsilon))\dot{h}_\varepsilon(t)\}\mathrm{d}t\\
&\quad + \sqrt{\varepsilon}\sigma(t, X^\varepsilon(t;\xi_\varepsilon,u_\varepsilon), X^\varepsilon(t-\tau;\xi_\varepsilon,u_\varepsilon))\mathrm{d}W(t)\\
&\quad + \int_{\mathbb{X}} \phi(t, X^\varepsilon(t-;\xi_\varepsilon,u_\varepsilon), X^\varepsilon((t-\tau)-;\xi_\varepsilon,u_\varepsilon), x)\\
&\quad \times (\varepsilon N^{\varepsilon^{-1}\varphi_\varepsilon}(\mathrm{d}t,\mathrm{d}x) - \nu_T(\mathrm{d}t,\mathrm{d}x))
\end{aligned} \tag{4.43}$$

with the initial data $X_0^\varepsilon(\xi_\varepsilon, u_\varepsilon) = \xi \in \mathscr{C}$. The following lemmas play a key role in completing the proof of Theorem 4.16.

Lemma 4.17 *Assume further that* (A1)–(A4) *hold. Then, there exists an* $\varepsilon_0 \in (0, 1)$ *such that*

$$\overline{\mathbb{E}}\Big(\sup_{0\le t\le T} |X^\varepsilon(t;\xi,u_\varepsilon)|^p\Big) < \infty,\ p \ge 2 \tag{4.44}$$

for any $u_\varepsilon = (h_\varepsilon, \varphi_\varepsilon) \in \overline{\mathscr{U}}_N$ *and* $\xi \in K$.

Proof Let $q' = q/2 + 1$ and set $M_\varepsilon(t) := X^\varepsilon(t;\xi,u_\varepsilon) - G(X^\varepsilon(t-\tau;\xi,u_\varepsilon))$. For arbitrary $p \ge 2$, from Remark 4.14, one has

$$|X^\varepsilon(t;\xi,u_\varepsilon)|^p \lesssim 1 + |M_\varepsilon(t)|^p + |X^\varepsilon(t-\tau;\xi,u_\varepsilon)|^{pq'}. \tag{4.45}$$

By the Itô formula, Remark 4.14, and (4.45), it follows that

$$\begin{aligned}
&\sup_{0\le s\le t} |X^\varepsilon(s;\xi,u_\varepsilon)|^2\\
&\lesssim 1 + \sup_{-\tau\le s\le t-\tau} |X^\varepsilon(s;\xi,u_\varepsilon)|^{2q'} + M_{1,\varepsilon}(t) + M_{2,\varepsilon}(t) + M_{3,\varepsilon}(t) \qquad (4.46)\\
&+ \int_0^t \Big\{(1 + |X^\varepsilon(s;\xi,u_\varepsilon)|^2 + |X^\varepsilon(s-\tau;\xi,u_\varepsilon)|^{2q'})\\
&\times \Big(1 + |\dot h_\varepsilon(s)|^2 + \int_{\mathbb{X}} \|x\|_{\mathbb{X}} \cdot |1 - \varphi_\varepsilon(s,x)|\nu(\mathrm{d}x)\Big)\Big\}\mathrm{d}s,
\end{aligned}$$

where

$$\begin{aligned}
M_{1,\varepsilon}(t) &:= 2\sqrt{\varepsilon}\sup_{0\le s\le t}\Big|\int_0^s \langle M_\varepsilon(z), \sigma(z, X^\varepsilon(z;\xi,u_\varepsilon), X^\varepsilon(z-\tau;\xi,u_\varepsilon))\mathrm{d}W(z)\rangle\Big|,\\
M_{2,\varepsilon}(t) &:= 2\varepsilon \sup_{0\le s\le t}\Big|\int_0^s\int_{\mathbb{X}} \langle M_\varepsilon(z-), \phi(z, X^\varepsilon(z-;\xi,u_\varepsilon), X^\varepsilon((z-\tau)-;\xi,u_\varepsilon), x)\rangle\\
&\quad \times (N^{\varepsilon^{-1}\varphi_\varepsilon}(\mathrm{d}z,\mathrm{d}x) - \varepsilon^{-1}\varphi_\varepsilon(z,x)\nu(\mathrm{d}x)\mathrm{d}z)\Big|,
\end{aligned}$$

and

$$M_{3,\varepsilon}(t) := \varepsilon^2 \int_0^t\int_{\mathbb{X}} |\phi(s, X^\varepsilon(s-;\xi,u_\varepsilon), X^\varepsilon((s-\tau)-;\xi,u_\varepsilon), x)|^2 N^{\varepsilon^{-1}\varphi_\varepsilon}(\mathrm{d}s,\mathrm{d}x).$$

In (4.46), making use of (A4) and [24, (3.4), Lemma 3.4] with $\Gamma_0(s,x) = \|x\|_{\mathbb{X}}$ and utilizing the Gronwall inequality yields

$$\sup_{0\le s\le t}|X^\varepsilon(s;\xi,u_\varepsilon)|^2 \lesssim 1 + \sup_{-\tau\le s\le t-\tau}|X^\varepsilon(s;\xi,u_\varepsilon)|^{2q'} + M_{1,\varepsilon}(t) + M_{2,\varepsilon}(t) + M_{3,\varepsilon}(t).$$

Hence, for any $p \geq 2$, we have

$$\overline{\mathbb{E}}\Big(\sup_{0\le s\le t}|X^{\varepsilon}(s;\xi,u_\varepsilon)|^p\Big) \le c\Big\{1+\overline{\mathbb{E}}\Big(\sup_{-\tau\le s\le t-\tau}|X^{\varepsilon}(s;\xi,u_\varepsilon)|^{pq'}\Big) + \overline{\mathbb{E}}M_{1,\varepsilon}^{p/2}(t)+\overline{\mathbb{E}}M_{2,\varepsilon}^{p/2}(t)+\overline{\mathbb{E}}M_{3,\varepsilon}^{p/2}(t)\Big\}. \tag{4.47}$$

Exploiting the B–D–G inequality and taking Remark 4.14 into account, we arrive at

$$\begin{aligned}
&\overline{\mathbb{E}}M_{1,\varepsilon}^{p/2}(t)+\overline{\mathbb{E}}M_{2,\varepsilon}^{p/2}(t)+\overline{\mathbb{E}}M_{3,\varepsilon}^{p/2}(t)\\
&\lesssim \varepsilon^{p/4}\Big\{1+\overline{\mathbb{E}}\Big(\sup_{0\le s\le t}|X^{\varepsilon,u_\varepsilon}(s)|^p\Big)+\overline{\mathbb{E}}\Big(\sup_{-\tau\le s\le t-\tau}|X^{\varepsilon,u_\varepsilon}(s)|^{pq'}\Big)\Big\}\\
&+\varepsilon^{p/4\wedge(p/2-1)}\Big\{1+\overline{\mathbb{E}}\Big(\sup_{0\le s\le t}|X^{\varepsilon,u_\varepsilon}(s)|^p\Big)+\overline{\mathbb{E}}\Big(\sup_{-\tau\le s\le t-\tau}|X^{\varepsilon,u_\varepsilon}(s)|^{pq'}\Big)\Big\} \qquad (4.48)\\
&\times\Big\{\Big(\int_0^t\int_{\mathbb{X}}\|x\|_{\mathbb{X}}^2\varphi_\varepsilon(s,x)\nu(\mathrm{d}x)\mathrm{d}s\Big)^{p/4}+\int_0^t\int_{\mathbb{X}}\|x\|_{\mathbb{X}}^p\varphi_\varepsilon(s,x)\nu(\mathrm{d}x)\mathrm{d}s\\
&+\Big(\int_0^t\int_{\mathbb{X}}\|x\|_{\mathbb{X}}^4\varphi_\varepsilon(s,x)\nu(\mathrm{d}x)\mathrm{d}s\Big)^{p/4}\Big\}.
\end{aligned}$$

Thus, according to (A4), [24, (3.4), Lemma 3.4] with $\Gamma_0(s,x)=\|x\|_{\mathbb{X}}^p, p\geq 2$, and $\varepsilon>0$ sufficiently small, we derive from (4.47) and (4.48) that

$$\overline{\mathbb{E}}\Big(\sup_{0\le s\le t}|X^{\varepsilon,u_\varepsilon}(s)|^p\Big)\lesssim 1+\overline{\mathbb{E}}\Big(\sup_{-\tau\le s\le t-\tau}|X^{\varepsilon,u_\varepsilon}(s)|^{pq'}\Big).$$

Finally, the desired assertion (4.44) follows from an induction argument.

Lemma 4.18 *Let* (A1)–(A4) *hold and assume further that* $u_n=(h_n,\varphi_n)\in S_N$, $u=(h,\varphi)\in S_N$ *and* $\xi_n,\xi\in K$ *such that* $u_n\to u$ *and* $\xi_n\to\xi$ *as* $n\to\infty$. *Then,*

$$\mathscr{G}^0\Big(\xi_n,\int_0^{\cdot}\dot h_n(s)\mathrm{d}s,\nu_T^{\varphi_n}\Big)\to\mathscr{G}^0\Big(\xi,\int_0^{\cdot}\dot h(s)\mathrm{d}s,\nu_T^{\varphi}\Big)\quad in\ C([0,T];\mathbb{R}^n),$$

where $\mathscr{G}^0$ *is introduced in* (4.40).

Proof By a close scrutiny of Lemma 4.10, to complete the proof, it suffices to show that

(i) $\sup_n\sup_{0\le t\le T}|Y^{u_n}(t;\xi)|<\infty$;
(ii) $\lim_{\delta\to 0}\sup_n|Y^{u_n}(t+\delta;\xi)-Y^{u_n}(t;\xi)|=0$;
(iii) $\lim_{n\to\infty}\sup_{0\le t\le T}|Y^{u_n}(t;\xi_n)-Y^{u_n}(t;\xi)|=0$.

Following an argument of Lemma 4.17, we deduce that (i) holds. Due to (i), (ii) holds by following the lines of Lemma 4.10. In what follows, we need only show that (iii) is also true. Set

$$Z^n(t) := Y^{u_n}(t;\xi_n) - Y^{u_n}(t;\xi)$$

and

$$M^n(t) := Y^{u_n}(t;\xi_n) - Y^{u_n}(t;\xi) - \{G(Y^{u_n}(t-\tau;\xi_n)) - G(Y^{u_n}(t-\tau;\xi))\}.$$

It follows from (A1) and (ii) that

$$\begin{aligned}|Z^n(t)|^2 &\lesssim |M^n(t)|^2 + (1+|Y^{u_n}(t-\tau;\xi)|^q + |Y^{u_n}(t-\tau;\xi_n)|^q)|Z^n(t-\tau)|^2\\ &\lesssim |M^n(t)|^2 + |Z^n(t-\tau)|^2.\end{aligned}$$

This, combining the chain rule with (A2)–(A3) and (i), gives that

$$\begin{aligned}|Z^n(t)|^2 &\lesssim |M^n(0)|^2 + |Z^n(t-\tau)|^2 + \int_0^t \Big\{(|Z^n(s)|^2\\ &\quad + (1+|Y^{u_n}(s-\tau;\xi)|^q + |Y^{u_n}(s-\tau;\xi_n)|^q)\\ &\quad \times |Z^n(s-\tau)|^2)\Big(1+|\dot{h}_n(s)| + \int_{\mathbb{X}} \|x\|_{\mathbb{X}} \cdot |1-\varphi_n(s,x)|\nu(\mathrm{d}x)\Big)\Big\}\mathrm{d}s \qquad (4.49)\\ &\lesssim |M^n(0)|^2 + |Z^n(t-\tau)|^2 + \int_0^t \Big\{(|Z^n(s)|^2 + |Z^n(s-\tau)|^2)\\ &\quad \times \Big(1+|\dot{h}_n(s)|^2 + \int_{\mathbb{X}} \|x\|_{\mathbb{X}} \cdot |1-\varphi_n(s,x)|\nu(\mathrm{d}x)\Big)\Big\}\mathrm{d}s.\end{aligned}$$

Next, according to $h_n \in S_{1,N}$, (A4) and [24, (3.4), Lemma 3.4] with $\Gamma_0(s,x) = \|x\|_{\mathbb{X}}$, we deduce from the Gronwall inequality that

$$\sup_{0\le s\le t} |Z^n(s)|^2 \lesssim |M^n(0)|^2 + \sup_{-\tau\le s\le t-\tau} |Z^n(s)|^2.$$

Consequently, (iii) follows from an induction argument.

Lemma 4.19 *Let* $u_\varepsilon = (h_\varepsilon, \varphi_\varepsilon) \in \overline{\mathscr{U}}_N$ *and* $\xi_\varepsilon, \xi \in K$ *such that* $\xi_\varepsilon \to \xi$ *as* $\varepsilon \to 0$, *and assume further that* (A1)–(A4) *hold. Then,*

$$\mathscr{G}^\varepsilon(\xi_\varepsilon, \sqrt{\varepsilon}W + \int_0^\cdot \dot{h}_\varepsilon(s)\mathrm{d}s, \varepsilon N^{\varepsilon^{-1}\varphi_\varepsilon}) - \mathscr{G}^\varepsilon(\xi, \sqrt{\varepsilon}W + \int_0^\cdot \dot{h}_\varepsilon(s)\mathrm{d}s, \varepsilon N^{\varepsilon^{-1}\varphi_\varepsilon}) \Rightarrow 0,$$

under $\bar{\mathbb{P}}$ *as* $\varepsilon \to 0$.

Proof Set

$$Z^\varepsilon(t) := X^\varepsilon(t;\xi_\varepsilon, u_\varepsilon) - X^\varepsilon(t;\xi, u_\varepsilon)$$

and

$$M^\varepsilon(t) := X^\varepsilon(t;\xi_\varepsilon,u_\varepsilon) - X^\varepsilon(t;\xi,u_\varepsilon) - \{G(X^\varepsilon(t-\tau;\xi_\varepsilon,u_\varepsilon)) - G(X^\varepsilon(t-\tau;\xi,u_\varepsilon))\}.$$

By the Itô formula, and (A1) and (A2), it follows that

$$\begin{aligned}
&|Z^\varepsilon(t)|^2 \\
&\lesssim |M^\varepsilon(0)|^2 + (1+|X^\varepsilon(t-\tau;\xi_\varepsilon)|^q + |X^\varepsilon(t-\tau;\xi)|^q)|Z^\varepsilon(t-\tau)|^2 \\
&+ \int_0^t \left\{|Z^\varepsilon(s)|^2 + (1+|X^\varepsilon(s-\tau;\xi_\varepsilon)|^q + |X^\varepsilon(s-\tau;\xi)|^q)|Z^\varepsilon(s-\tau)|^2\right\} \\
&\left(1+|\dot{h}_\varepsilon(s)|^2 + \int_{\mathbb{X}} \|x\|_{\mathbb{X}}\cdot|1-\varphi_\varepsilon(s,x)|\nu(\mathrm{d}x)\right)\mathrm{d}s \qquad (4.50)\\
&+\varepsilon^2\int_0^t\int_{\mathbb{X}} |\phi(s,X^\varepsilon(s-;\xi_\varepsilon,u_\varepsilon),X^\varepsilon((s-\tau)-;\xi_\varepsilon,u_\varepsilon),x) \\
&-\phi(s,X^\varepsilon(s-;\xi,u_\varepsilon),X^\varepsilon((s-\tau)-;\xi,u_\varepsilon),x)|^2 N^{\varepsilon^{-1}\varphi_\varepsilon}(\mathrm{d}s,\mathrm{d}x) \\
&+M_{1,\varepsilon}(t)+M_{2,\varepsilon}(t),
\end{aligned}$$

where

$$\begin{aligned}
M_{1,\varepsilon}(t) :&= 2\sqrt{\varepsilon}\sup_{0\le s\le t}\left|\int_0^s \langle M^\varepsilon(r), (\sigma(r,X^\varepsilon(r;\xi_\varepsilon,u_\varepsilon),X^\varepsilon(r-\tau;\xi_\varepsilon,u_\varepsilon))\right.\\
&\left.-\sigma(r,X^\varepsilon(r;\xi,u_\varepsilon),X^\varepsilon(r-\tau;\xi,u_\varepsilon)))\mathrm{d}W(r)\rangle\right|,
\end{aligned}$$

and

$$\begin{aligned}
M_{2,\varepsilon}(t) :&= 2\varepsilon\sup_{0\le s\le t}\left|\int_0^s\int_{\mathbb{X}} \langle M^\varepsilon(r-), \phi(r,X^\varepsilon(r-;\xi_\varepsilon,u_\varepsilon),X^\varepsilon((r-\tau)-;\xi_\varepsilon,u_\varepsilon),x)\right.\\
&-\phi(r,X^\varepsilon(r-;\xi,u_\varepsilon),X^\varepsilon((r-\tau)-;\xi,u_\varepsilon),x)\rangle\\
&\left.\times(N^{\varepsilon^{-1}\varphi_\varepsilon}(\mathrm{d}r,\mathrm{d}x)-\varepsilon^{-1}\varphi_\varepsilon(r,x)\nu(\mathrm{d}x)\mathrm{d}r)\right|.
\end{aligned}$$

Let

$$J(T) = 1 + \sup_{-\tau\le t\le T}|X^\varepsilon(t-\tau;\xi_\varepsilon)|^q + \sup_{-\tau\le t\le T}|X^\varepsilon(t;\xi)|^q.$$

Using the Gronwall inequality, (A4), and [24, (3.4), Lemma 3.4] with $\Gamma_0(s,x) = \|x\|_{\mathbb{X}}$, we derive from (4.50) that

$$
\begin{aligned}
I(t) := \sup_{0\le s\le t} |Z^\varepsilon(s)|^2 &\le |M^\varepsilon(0)|^2 + J(T) \sup_{-\tau\le s\le t-\tau} |Z^\varepsilon(s)|^2 \\
&\quad + \varepsilon^2 \int_0^t \int_{\mathbb{X}} |\phi(s, X^\varepsilon(s-;\xi_\varepsilon, u_\varepsilon), X^\varepsilon((s-\tau)-;\xi_\varepsilon, u_\varepsilon), x) \\
&\quad - \phi(s, X^\varepsilon(s-;\xi, u_\varepsilon), X^\varepsilon((s-\tau)-;\xi, u_\varepsilon), x)|^2 N^{\varepsilon^{-1}\varphi_\varepsilon}(\mathrm{d}r, \mathrm{d}x) \\
&\quad + M_{1,\varepsilon}(t) + M_{2,\varepsilon}(t) \\
&=: |M^\varepsilon(0)|^2 + \sum_{i=1}^4 I_i(t).
\end{aligned}
$$

For any $\delta > 0$, it is readily seen that

$$
\overline{\mathbb{P}}\Big(I(t) \ge \delta\Big) \le \overline{\mathbb{P}}(|M^\varepsilon(0)|^2 \ge \delta/5) + \sum_{i=1}^4 \overline{\mathbb{P}}\Big(I_i(t) \ge \delta/5\Big).
$$

On the other hand, for any $R > 0$, Chebyshev's inequality and (4.44) yield that

$$
\begin{aligned}
\overline{\mathbb{P}}\Big(I_1(t) \ge \delta/4\Big) &\le \frac{1}{R}\overline{\mathbb{E}}J(T) + \overline{\mathbb{P}}\Big(I(t-\tau) \ge \delta/(5R)\Big) \\
&\lesssim \frac{1}{R} + \overline{\mathbb{P}}\Big(I(t-\tau) \ge \delta/(5R)\Big).
\end{aligned}
$$

Furthermore, by following an argument of (4.22), we deduce from (4.44) that

$$
\sum_{i=2}^4 \overline{\mathbb{P}}\Big(I_i(t) \ge \delta/5\Big) \lesssim \sqrt{\varepsilon}.
$$

Hence, we arrive at

$$
\overline{\mathbb{P}}\Big(I(t) \ge \delta\Big) \lesssim \sqrt{\varepsilon} + \overline{\mathbb{P}}(|M^\varepsilon(0)|^2 \ge \delta/5) + \frac{1}{R} + \overline{\mathbb{P}}\Big(I(t-\tau) \ge \delta/(5R)\Big).
$$

Then, the desired assertion follows from an induction argument and by taking $R > 0$ sufficiently large.

Lemma 4.20 *Assume that $u_\varepsilon = (h_\varepsilon, \varphi_\varepsilon) \subset \overline{\mathcal{U}}_N$, $u = (h, \varphi) \in \overline{\mathcal{U}}_N$ and $\xi_\varepsilon, \xi \in K$ such that $u_\varepsilon \Rightarrow u$, under $\overline{\mathbb{P}}$, and $\xi_\varepsilon \to \xi$ as $\varepsilon \to 0$. Then, under $\overline{\mathbb{P}}$, as $\varepsilon \to 0$,*

$$
\mathcal{G}^\varepsilon(\xi_\varepsilon, \sqrt{\varepsilon}W + \int_0^\cdot \dot{h}_\varepsilon(s)\mathrm{d}s, \varepsilon N^{\varepsilon^{-1}\varphi_\varepsilon}) \Rightarrow \mathcal{G}^0\Big(\xi, \int_0^\cdot \dot{h}(s)\mathrm{d}s, \nu_T^\varphi\Big).
$$

Proof By virtue of Lemma 4.19, we only show that $(X^\varepsilon(\cdot;\xi, u_\varepsilon))_{\varepsilon>0}$ is tight in $D([0,T];\mathbb{R}^n)$ since the rest of proof follows the same lines of Lemma 4.12. Using (4.44), $(X^\varepsilon(\cdot;\xi, u_\varepsilon))_{\varepsilon>0}$ is tight on the line for each $t \in [0, T]$. Let

$$M_\varepsilon(\tau_\varepsilon + \delta_\varepsilon) = X^\varepsilon(\tau_\varepsilon + \delta_\varepsilon; \xi, u_\varepsilon) - G(X^\varepsilon(\tau_\varepsilon + \delta_\varepsilon - \tau; \xi, u_\varepsilon)).$$

For any $\{\tau_\varepsilon, \delta_\varepsilon\}$ satisfying (i) and (ii) in Lemma 4.11, from (A1) it follows that

$$\begin{aligned} &|X^\varepsilon(\tau_\varepsilon + \delta_\varepsilon; \xi, u_\varepsilon) - X^\varepsilon(\tau_\varepsilon; \xi, u_\varepsilon)| \\ &\le |M_\varepsilon(\tau_\varepsilon + \delta_\varepsilon) - M_\varepsilon(\tau_\varepsilon)| \\ &\quad + \Lambda_\varepsilon |X^\varepsilon(\tau_\varepsilon + \delta_\varepsilon - \tau; \xi, u_\varepsilon) - X^\varepsilon(\tau_\varepsilon - \tau; \xi, u_\varepsilon)|, \end{aligned} \tag{4.51}$$

in which

$$\Lambda_\varepsilon := c(1 + |X^\varepsilon(\tau_\varepsilon + \delta_\varepsilon - \tau; \xi, u_\varepsilon)|^{q/2} + |X^\varepsilon(\tau_\varepsilon - \tau; \xi, u_\varepsilon)|^{q/2})$$

for some $c > 0$. Following an argument of (4.36), we conclude that

$$|M_\varepsilon(\tau_\varepsilon + \delta_\varepsilon) - M_\varepsilon(\tau_\varepsilon)| \to 0 \quad \text{in probability as } \varepsilon \to 0. \tag{4.52}$$

Moreover, for any $\delta, R > 0$, the Chebyshev inequality leads to

$$\begin{aligned} &\overline{\mathbb{P}}(\Lambda_\varepsilon |X^\varepsilon(\tau_\varepsilon + \delta_\varepsilon - \tau; \xi, u_\varepsilon) - X^\varepsilon(\tau_\varepsilon - \tau; \xi, u_\varepsilon)| \ge \delta/2) \\ &\le \overline{\mathbb{P}}(\Lambda_\varepsilon \ge R) + \overline{\mathbb{P}}(|X^\varepsilon(\tau_\varepsilon + \delta_\varepsilon - \tau; \xi, u_\varepsilon) - X^\varepsilon(\tau_\varepsilon - \tau; \xi, u_\varepsilon)| \ge \delta/(2R)) \\ &\le \frac{1}{R}\overline{\mathbb{E}}\Lambda_\varepsilon + \overline{\mathbb{P}}(|X^\varepsilon(\tau_\varepsilon + \delta_\varepsilon - \tau; \xi, u_\varepsilon) - X^\varepsilon(\tau_\varepsilon - \tau; \xi, u_\varepsilon)| \ge \delta/(2R)). \end{aligned} \tag{4.53}$$

Thus, for any $\delta > 0$, combining (4.51) with (4.44) and (4.53) gives that

$$\begin{aligned} &\overline{\mathbb{P}}(|X^\varepsilon(\tau_\varepsilon + \delta_\varepsilon - \tau; \xi, u_\varepsilon) - X^\varepsilon(\tau_\varepsilon - \tau; \xi, u_\varepsilon)| \ge \delta) \\ &\le \overline{\mathbb{P}}(|M_\varepsilon(\tau_\varepsilon + \delta_\varepsilon) - M_\varepsilon(\tau_\varepsilon)| \ge \delta/2) + \frac{c}{R} \\ &\quad + \overline{\mathbb{P}}(|X^\varepsilon(\tau_\varepsilon + \delta_\varepsilon - \tau; \xi, u_\varepsilon) - X^\varepsilon(\tau_\varepsilon - \tau; \xi, u_\varepsilon)| \ge \delta/(2R)). \end{aligned}$$

By taking $R > 0$ sufficiently large and recalling (4.52), (II) in Lemma 4.11 follows from an induction argument. Hence, according to Lemma 4.11, $(X^\varepsilon(\cdot; \xi, u_\varepsilon))_{\varepsilon>0}$ is tight in $D([0, T]; \mathbb{R}^n)$.

Proof of Theorem 4.16

Proof According to Lemma 4.5, the assertion of Theorem 4.16 follows from Lemmas 4.18 and 4.20 immediately.

4.5 Moderate Deviations for SDDEs with Jumps

In this section, we focus on an FSDE taking the form

$$dX^\varepsilon(t) = b(X_t^\varepsilon)dt + \varepsilon \int_{\mathbb{X}} \phi(X_{t-}^\varepsilon, x) N^{\varepsilon^{-1}}(dt, dx), \quad t > 0 \tag{4.54}$$

with the initial value $X_0^\varepsilon = \xi \in \mathscr{D}$, where $b : \mathscr{D} \mapsto \mathbb{R}^n$, and $\phi : \mathscr{D} \times \mathbb{X} \mapsto \mathbb{R}^n$. For the present setup, we have $\overline{\mathbb{V}} = \overline{\mathbb{M}}$, and we point out that the Poisson measure $N^{\varepsilon^{-1}}$, and the notation $\overline{\mathbb{P}}$, $\overline{\mathscr{A}}_b$ below are defined similarly as in Section 4.2.

As $\varepsilon \to 0$, the solution X^ε to (4.54) tends to X^0 which solves the following deterministic functional differential equation

$$dX^0(t) = \Big(b(X_t^0) + \int_{\mathbb{X}} \phi(X_t^0, x)\nu(dx)\Big)dt, \quad t > 0 \tag{4.55}$$

with the initial value $X_0^0 = \xi \in \mathscr{D}$. In this section, we are interested in an MDP for X^ε, which corresponds to the LDP of the trajectory

$$Y^\varepsilon(t) := \frac{1}{a(\varepsilon)}(X^\varepsilon(t) - X^0(t)), \qquad t \in [0, T], \tag{4.56}$$

where $a(\varepsilon)$ satisfies

$$a(\varepsilon) \to 0 \text{ and } \beta(\varepsilon) := \frac{\varepsilon}{a^2(\varepsilon)} \to 0 \text{ as } \varepsilon \to 0. \tag{4.57}$$

It is worth pointing out that for simplicity we consider only the present setting (i.e., (4.54)), where the noise is given in terms of a Poisson random measure while there is no neutral term and Brownian motion noise. However, the more general framework, where the neutral term, Poisson, and Brownian motion noises are present, can be treated similarly.

The theory of MDPs is concerned with probabilities of deviations with a smaller order than in the large deviation theory. For the study on MDPs, we refer to [31, 38] for independent random vectors, and [147] for dependent random variables, [25, 140, 141] for stochastic dynamical systems in finite and infinite dimensions. Before proceeding further, we recall some additional notation and preliminary results. For further details, we refer the reader to [25]. The notion below concerns LDPs with general speed.

Definition 4.21 Given a collection $(\beta(\varepsilon))_{\varepsilon>0}$ of positive real numbers, the sequence $(Y^\varepsilon)_{\varepsilon>0}$ is said to satisfy an LDP on $\mathbb{S}$ with speed $\beta(\varepsilon)$ and rate function I, if for each closed subset F of $\mathbb{S}$,

$$\limsup_{\varepsilon\to 0}(\beta(\varepsilon) \log \mathbb{P}(Y^\varepsilon \in F)) \le - \inf_{f\in F}\{I(f)\},$$

and, for each open subset G of $\mathbb{S}$,

$$\liminf_{\varepsilon\to 0}(\beta(\varepsilon)\log\mathbb{P}(Y^\varepsilon\in G))\geq -\inf_{f\in G}\{I(f)\}.$$

For any $\varepsilon>0$ and $M\in(0,\infty)$, set

$$S^M_{+,\varepsilon}:=\{\varphi:\mathbb{X}_T\mapsto\mathbb{R}_+|L_{2,T}(\varphi)\leq Ma^2(\varepsilon)\},$$

where $L_{2,T}(\cdot)$ was defined in (4.5), and

$$S^M_\varepsilon:=\{\psi:\mathbb{X}_T\mapsto\mathbb{R}|\psi=(\varphi-1)/a(\varepsilon),\varphi\in S^M_{+,\varepsilon}\}.$$

Let

$$\mathscr{U}^M_{+,\varepsilon}=\{\varphi\in\overline{\mathscr{A}}_b:\varphi(\cdot,\cdot,\omega)\in S^M_{+,\varepsilon},\overline{\mathbb{P}}-\text{a.s.}\},$$

and

$$\mathscr{U}^M_\varepsilon=\{\varphi\in\overline{\mathscr{A}}_2:\varphi(\cdot,\cdot,\omega)\in S^M_\varepsilon,\overline{\mathbb{P}}-\text{a.s.}\},$$

where $\overline{\mathscr{A}}_2$ is introduced in (4.2). Let

$$L^2(\nu_T)=\left\{f:[0,T]\times\mathbb{X}\mapsto\mathbb{R}\middle|\|f\|_2:=\left(\int_{\mathbb{X}_T}|f(t,x)|^2\nu_T(\mathrm{d}t,\mathrm{d}x)\right)^{1/2}<\infty\right\},$$

which is a Hilbert space endowed with the norm $\|\cdot\|_2$. The ball of radius r centered at 0 in $L^2(\nu_T)$ is denoted by $B_2(r):=\{f\in L^2(\nu_T):\|f\|_2\leq r\}$. By [25, Lemma 3.2 (c)], there exists $\kappa_2(1)$ such that

$$\psi 1_{\{|\psi|\leq 1/a(\varepsilon)\}}\in B_2((M\kappa_2(1))^{1/2}),$$

where $\psi:=(\varphi-1)/a(\varepsilon)$, $\varphi\in S^M_{+,\varepsilon}$.

For some measurable maps $\mathscr{G}^0:L^2(\nu_T)\mapsto\mathbb{S}$ and $\mathscr{G}^\varepsilon:\overline{\mathbb{M}}\mapsto\mathbb{S}$, we assume that the following two conditions hold.

(a) Given an $M\in(0,\infty)$, suppose that $g^\varepsilon, g\in B_2(M)$ and $g^\varepsilon\to g$. Then

$$\mathscr{G}^0(g^\varepsilon)\to\mathscr{G}^0(g);$$

(b) Given an $M\in(0,\infty)$, let $(\varphi^\varepsilon)_{\varepsilon>0}$ be such that for every $\varepsilon>0$, $\varphi^\varepsilon\in\mathscr{U}^M_{+,\varepsilon}$ and $\psi^\varepsilon 1_{\{|\psi^\varepsilon|\leq 1/a(\varepsilon)\}}\Rightarrow\psi$ in $B_2((M\kappa_2(1))^{1/2})$ with $\psi^\varepsilon:=(\varphi^\varepsilon-1)/a(\varepsilon)$. Then

$$\mathscr{G}^\varepsilon(\varepsilon N^{\varepsilon^{-1}\varphi_\varepsilon})\Rightarrow\mathscr{G}^0(\psi).$$

The criteria below on LDPs with general speed is taken from [25, Theorem 2.3].

Lemma 4.22 *Suppose that the maps $\mathscr{G}^\varepsilon$ and $\mathscr{G}^0$ satisfy* (a) *and* (b) *above. Then*

$$I(\eta) = \inf_{\psi \in \mathbb{S}_\eta} \left[\frac{1}{2} \|\psi\|_2^2 \right], \tag{4.58}$$

where

$$\mathbb{S}_\eta = \mathbb{S}_\eta(\mathscr{G}^0) = \{\psi \in L^2(\nu_T) : \eta = \mathscr{G}^0(\psi)\},$$

is a rate function and ($Y^\varepsilon = \mathscr{G}^\varepsilon(\varepsilon N^{\varepsilon^{-1}})$) *satisfies an LDP with speed* $\beta(\varepsilon)$ *and rate function* I.

For any $x \in \mathbb{X}$, set

$$\Gamma_0(x) := \sup_{\xi \in \mathscr{D}} \frac{\phi(\xi, x)}{1 + \|\xi\|_\infty} \quad \text{and} \quad \Gamma_1(x) := \sup_{\xi, \eta \in \mathscr{D}, \xi \neq \eta} \frac{\phi(\xi, x) - \phi(\eta, x)}{\|\xi - \eta\|_\infty}.$$

Assume that

(C1) $b : \mathscr{D} \mapsto \mathbb{R}^n$ is differentiable, $\phi : \mathscr{D} \times \mathbb{X} \mapsto \mathbb{R}^n$ is differentiable with respect to the first variable, and there exists an $L > 0$ such that for any $\xi, \eta \in \mathscr{D}$,

$$|b(\xi) - b(\eta)| + \int_{\mathbb{X}} |\phi(\xi, x) - \phi(\eta, x)| \nu(\mathrm{d}x) \le L \|\xi - \eta\|_\infty. \tag{4.59}$$

(C2) $\Gamma_0, \Gamma_1 \in L^1(\nu) \cap L^2(\nu)$ and there exist $\delta_1, \delta_2 \in (0, \infty)$ such that for $\forall\, E \in \mathbb{X}$ with $\nu(E) < \infty$,

$$\int_E \exp(\delta_1 \Gamma_0^2(x)) \nu(\mathrm{d}x) < \infty \quad \text{and} \quad \int_E \exp(\delta_2 \Gamma_1^2(x)) \nu(\mathrm{d}x) < \infty.$$

(C3) There exists an $L_0 > 0$ such that for any $\xi, \eta \in \mathscr{D}$,

$$\|\nabla b(\xi) - \nabla b(\eta)\| + \int_{\mathbb{X}} \|\nabla \phi(\xi, x) - \nabla \phi(\eta, x)\| \nu(\mathrm{d}x) \le L_0 \|\xi - \eta\|_\infty.$$

Our main result in this section is stated next.

Theorem 4.23 *Assume that* (C1)–(C3) *hold. Then,* (Y^ε), *defined by* (4.56), *satisfies an LDP in* $D([0, T]; \mathbb{R}^n)$ *with speed* $\beta(\varepsilon)$ *and the rate function I, defined by* (4.58), *wherein*

$$\mathrm{d}\eta(t) = \left(\nabla_{\eta_t} b(X_t^0) + \int_{\mathbb{X}} \nabla_{\eta_t} \phi(X_t^0, x) \nu(\mathrm{d}x) + \int_{\mathbb{X}} \psi(t, x) \phi(X_t^0, x) \nu(\mathrm{d}x) \right) \mathrm{d}t \tag{4.60}$$

with the initial value $\eta_0 \equiv 0$, *where* $\psi \in L^2(\nu_T)$.

To highlight the dependence on ψ of the solution η to (4.60), in what follows we write η^ψ instead of η. Define

$$\mathscr{G}^0(\psi) = \eta^\psi, \quad \psi \in L^2(\nu_T), \tag{4.61}$$

where η^ψ solves (4.60). The proof of Theorem 4.23 is based on the following lemmas.

Lemma 4.24 *Assume that* (4.59) *and* $\Gamma_0 \in L^1(\nu)$ *hold. Then there exists some* $C_T > 0$ *such that*

$$\sup_{t\in[0,T]} \|X_t^0\|_\infty \le C_T. \tag{4.62}$$

Proof By the chain rule, it follows from (4.59) and $\Gamma_0 \in L^1(\nu)$ that

$$\begin{aligned}|X^0(t)|^2 &\lesssim \|\xi\|_\infty^2 + \Big(1 + \int_{\mathbb{X}} \Gamma_0(x)\nu(\mathrm{d}x)\Big)\int_0^t (1 + \|X_s^0\|_\infty^2)\mathrm{d}s\\ &\lesssim \|\xi\|_\infty^2 + \int_0^t (1 + \|X_s^0\|_\infty^2)\mathrm{d}s, \quad t \in [0,T].\end{aligned}$$

This further gives that

$$1 + \sup_{0\le s\le t} \|X_s^0\|_\infty^2 \lesssim 1 + \|\xi\|_\infty^2 + \int_0^t \Big(1 + \sup_{0\le r\le s} \|X_r^0\|_\infty^2\Big)\mathrm{d}s, \quad t \in [0,T].$$

Thus, the desired assertion follows from the Gronwall inequality.

The lemma below aims to verify that the measurable map $\mathscr{G}^0$, defined by (4.61), satisfies (a) in Lemma 4.22.

Lemma 4.25 *Assume that* (4.59) *holds,* $\Gamma_0 \in L^1(\nu) \cap L^2(\nu)$, *and* $\Gamma_1 \in L^1(\nu)$, *and suppose further that* $g^\varepsilon \to g$ *whenever* $g^\varepsilon, g \in B_2(M)$ *for some* $M \in (0,\infty)$. *Then,* $\mathscr{G}^0(g^\varepsilon) \to \mathscr{G}^0(g)$.

Proof Let $Z^\varepsilon(t) = \eta^{g^\varepsilon}(t)$, which solves (4.60) with $\psi = g^\varepsilon$ therein. By following the lines of argument for Lemma 4.10, to end the proof of Lemma 4.25, it is sufficient to show that

(i) $(Z^\varepsilon(\cdot))_{\varepsilon\in(0,1)}$ is uniformly bounded, i.e., $\sup_{\varepsilon\in(0,1)} \sup_{t\in[0,T]} |Z^\varepsilon(t)| < \infty$;
(ii) $(Z^\varepsilon(\cdot))_{\varepsilon\in(0,1)}$ is uniformly equicontinuous, i.e.,

$$\lim_{\delta\to 0} \sup_{\varepsilon\in(0,1)} |Z^\varepsilon(t+\delta) - Z^\varepsilon(t)| = 0.$$

For any $\xi, \eta \in \mathscr{D}$, observe from (4.59) that

$$|\nabla_\eta b(\xi)| \le L\|\eta\|_\infty, \quad \int_{\mathbb{X}} |\nabla_\eta \phi(\xi, x)|\nu(\mathrm{d}x) \le \Big(L\int_{\mathbb{X}} \Gamma_1(x)\nu(\mathrm{d}x)\Big)\|\eta\|_\infty. \tag{4.63}$$

By the chain rule, we derive from (4.63) and Hölder's inequality that

$$
\begin{aligned}
|Z^\varepsilon(t)|^2 &\lesssim \left(1+\left(L\int_{\mathbb{X}}\Gamma_1(x)\nu(\mathrm{d}x)\right)^2\right)\int_0^t \|Z_s^\varepsilon\|_\infty^2\mathrm{d}s \\
&\quad+\int_0^t\int_{\mathbb{X}}|g^\varepsilon(s,x)|^2\nu(\mathrm{d}x)\int_{\mathbb{X}}\Gamma_0^2(x)\nu(\mathrm{d}x)(1+\|X_s^0\|_\infty^2)\mathrm{d}s \\
&\lesssim \int_0^t\int_{\mathbb{X}}|g^\varepsilon(s,x)|^2\nu(\mathrm{d}x)\mathrm{d}s+\int_0^t\|Z_s^\varepsilon\|_\infty^2\mathrm{d}s,
\end{aligned}
$$

where in the last display we have used $\Gamma_0\in L^2(\nu)$, $\Gamma_1\in L^1(\nu)$, and (4.62). As a result, (i) follows from Gronwall's inequality and $g^\varepsilon\in B_2(M)$.

Note from (4.62) and (4.63), in addition to $\Gamma_0\in L^2(\nu)$ and $\Gamma_1\in L^1(\nu)$, that

$$
\begin{aligned}
&|Z^\varepsilon(t+\delta)-Z^\varepsilon(t)| \\
&\lesssim\left(1+\int_{\mathbb{X}}\Gamma_1(x)\nu(\mathrm{d}x)\right)\int_t^{t+\delta}\|Z_s^\varepsilon\|_\infty\mathrm{d}s \\
&\quad+\int_t^{t+\delta}\int_{\mathbb{X}}|g^\varepsilon(s,x)|\Gamma_0(x)(1+\|X_s^0\|_\infty)\nu(\mathrm{d}x)\mathrm{d}s \\
&\lesssim\int_t^{t+\delta}\|Z_s^\varepsilon\|_\infty\mathrm{d}s+\int_t^{t+\delta}\left(\int_{\mathbb{X}}|g^\varepsilon(s,x)|^2\nu(\mathrm{d}x)\right)^{1/2}\left(\int_{\mathbb{X}}|\Gamma_0(x)|^2\nu(\mathrm{d}x)\right)^{1/2}\mathrm{d}s \\
&\lesssim\int_t^{t+\delta}\|Z_s^\varepsilon\|_\infty\mathrm{d}s+\int_t^{t+\delta}\left(\int_{\mathbb{X}}|g^\varepsilon(s,x)|^2\nu(\mathrm{d}x)\right)^{1/2}\mathrm{d}s,\quad \delta\in(0,1).
\end{aligned}
$$

This, together with (i), $g^\varepsilon\in B_2(M)$, and Hölder's inequality, yields (ii).

By Yamada–Watanabe's theorem, there is a measurable map $\overline{\mathscr{G}}^\varepsilon:\mathbb{M}\mapsto\mathbb{S}$ obeying $X^\varepsilon=\overline{\mathscr{G}}^\varepsilon(\varepsilon N^{\varepsilon^{-1}})$. Next, by the Girsanov theorem, for any $\varphi\in\mathscr{U}_{+,\varepsilon}^M$, $\overline{X}^{\varepsilon,\varphi}:=\overline{\mathscr{G}}^\varepsilon(\varepsilon N^{\varepsilon^{-1}\varphi})$ solves

$$
\mathrm{d}\overline{X}^{\varepsilon,\varphi}(t)=b(\overline{X}_t^{\varepsilon,\varphi})\mathrm{d}t+\varepsilon\int_{\mathbb{X}}\phi(\overline{X}_{t-}^{\varepsilon,\varphi},x)N^{\varepsilon^{-1}\varphi}(\mathrm{d}t,\mathrm{d}x),\quad t>0.
$$

So, there exists a measurable map $\mathscr{G}^\varepsilon:\overline{\mathbb{M}}\mapsto\mathbb{S}$ such that

$$
\overline{Y}^{\varepsilon,\varphi}:=\mathscr{G}^\varepsilon(\varepsilon N^{\varepsilon^{-1}\varphi})=\frac{1}{a(\varepsilon)}(\overline{X}^{\varepsilon,\varphi}-X^0). \tag{4.64}
$$

Lemma 4.26 *Assume that* (4.59) *and* (C2) *hold. Then, there exists an* $\varepsilon_0\in(0,1)$ *such that for some* $\overline{C}_T>0$,

$$
\mathbb{E}\Big(\sup_{0\le t\le T}\|\overline{Y}_t^{\varepsilon,\varphi}\|_\infty^2\Big)\le\overline{C}_T,\quad \varepsilon\in(0,\varepsilon_0),\quad \varphi\in\mathscr{U}_{+,\varepsilon}^M. \tag{4.65}
$$

Proof Note that $\overline{Y}^{\varepsilon,\varphi}$ can be decomposed into the following five parts:

$$\begin{aligned}\overline{Y}^{\varepsilon,\varphi}(t) &= \frac{\varepsilon}{a(\varepsilon)}\int_0^t\int_{\mathbb{X}}\phi(\overline{X}^{\varepsilon,\varphi}_{s-},x)\tilde{N}^{\varepsilon^{-1}\varphi}(\mathrm{d}s,\mathrm{d}x)\\&+\frac{1}{a(\varepsilon)}\int_0^t\{b(\overline{X}^{\varepsilon,\varphi}_s)-b(X^0_s)\}\mathrm{d}s\\&+\frac{1}{a(\varepsilon)}\int_0^t\int_{\mathbb{X}}\{\phi(\overline{X}^{\varepsilon,\varphi}_s,x)-\phi(X^0_s,x)\}\nu(\mathrm{d}x)\mathrm{d}s\\&+\int_0^t\int_{\mathbb{X}}\{\phi(\overline{X}^{\varepsilon,\varphi}_s,x)-\phi(X^0_s,x)\}\psi(s,x)\nu(\mathrm{d}x)\mathrm{d}s\\&+\int_0^t\int_{\mathbb{X}}\phi(X^0_s,x))\psi(s,x)\nu(\mathrm{d}x)\mathrm{d}s=:\sum_{i=1}^5 I_i^{\varepsilon,\varphi}(t),\end{aligned}\tag{4.66}$$

where $\psi=(\varphi-1)/a(\varepsilon)$. By the Doob martingale inequality, it follows from (C2), Lemma 4.9, and [25, Lemma 4.2] that

$$\begin{aligned}&\overline{\mathbb{E}}\Big(\sup_{0\le s\le t}|I_1^{\varepsilon,\varphi}(s)|^2\Big)\\&\lesssim\beta(\varepsilon)\overline{\mathbb{E}}\Big(\Big(1+\sup_{0\le t\le T}\|\overline{X}^{\varepsilon,\varphi}_t\|^2_\infty\Big)\int_0^t\int_{\mathbb{X}}\Gamma_0^2(x)\varphi(s,x)\nu(\mathrm{d}x)\mathrm{d}s\Big)\\&\lesssim\beta(\varepsilon).\end{aligned}\tag{4.67}$$

Applying the Hölder inequality and recalling $\Gamma_1\in L^1(\nu)$, we obtain from (4.59) that

$$\begin{aligned}\sum_{i=2}^3\overline{\mathbb{E}}\Big(\sup_{0\le s\le t}|I_2^{\varepsilon,\varphi}(s)|^2\Big)&\lesssim\int_0^t\overline{\mathbb{E}}\|\overline{Y}^{\varepsilon,\varphi}_s\|^2_\infty\mathrm{d}s+\overline{\mathbb{E}}\int_0^t\Big(\|\overline{Y}^{\varepsilon,\varphi}_s\|_\infty\int_{\mathbb{X}}\Gamma_1(x)\nu(\mathrm{d}x)\Big)^2\mathrm{d}s\\&\lesssim\int_0^t\overline{\mathbb{E}}\|\overline{Y}^{\varepsilon,\varphi}_s\|^2_\infty\mathrm{d}s.\end{aligned}$$

Next, taking (4.62) and [25, Lemma 4.3] into account yields that

$$\begin{aligned}&\sum_{i=4}^5\overline{\mathbb{E}}\Big(\sup_{0\le s\le t}|I_i^{\varepsilon,\varphi}(s)|^2\Big)\\&\le\overline{\mathbb{E}}\Big(\sup_{0\le s\le t}\|\overline{Y}^{\varepsilon,\varphi}_s\|^2_\infty\Big)a^2(\varepsilon)\Big(\int_0^t\int_{\mathbb{X}}\Gamma_1(x)\psi(s,x)\nu(\mathrm{d}x)\mathrm{d}s\Big)^2\\&\quad+\Big(1+\sup_{0\le s\le t}\|X^0_s\|_\infty\Big)^2\Big(\int_0^t\int_{\mathbb{X}}\Gamma_0(x)\psi(s,x)\nu(\mathrm{d}x)\mathrm{d}s\Big)^2\\&\lesssim 1+a^2(\varepsilon)\overline{\mathbb{E}}\Big(\sup_{0\le s\le t}\|\overline{Y}^{\varepsilon,\varphi}_s\|^2_\infty\Big).\end{aligned}\tag{4.68}$$

Thus, we arrive at

$$\overline{\mathbb{E}}\Big(\sup_{0\le s\le t}\|\overline{Y}_s^{\varepsilon,\varphi}\|_\infty^2\Big)\le \bar c\Big\{1+\beta(\varepsilon)+\|\xi\|_\infty^2+\int_0^t\overline{\mathbb{E}}\|\overline{Y}_s^{\varepsilon,\varphi}\|_\infty^2\mathrm{d}s$$
$$+a^2(\varepsilon)\overline{\mathbb{E}}\Big(\sup_{0\le s\le t}\|\overline{Y}_s^{\varepsilon,\varphi}\|_\infty^2\Big)\Big\}$$

for some $\bar c>0$. Taking $\varepsilon>0$ sufficiently small such that $\bar c a^2(\varepsilon)\le\frac12$ leads to

$$\overline{\mathbb{E}}\Big(\sup_{0\le s\le t}\|\overline{Y}_s^{\varepsilon,\varphi}\|_\infty^2\Big)\lesssim 1+\beta(\varepsilon)+\|\xi\|_\infty^2+\int_0^t\overline{\mathbb{E}}\Big(\sup_{0\le r\le s}\|\overline{Y}_r^{\varepsilon,\varphi}\|_\infty^2\Big)\mathrm{d}s.$$

Then, the desired assertion follows from Gronwall's inequality and by noting that $\beta(\varepsilon)\to0$ as $\varepsilon\to0$.

The next lemma shows that $\mathscr{G}^\varepsilon$, defined by (4.64) satisfies (b) in Lemma 4.22.

Lemma 4.27 *Given an $M\in(0,\infty)$, let $(\varphi^\varepsilon)_{\varepsilon>0}$ be such that for every $\varepsilon>0$, $\varphi^\varepsilon\in\mathscr{U}^M_{+,\varepsilon}$, $\psi^\varepsilon 1_{\{|\psi^\varepsilon|\le1/a(\varepsilon)\}}\Rightarrow\psi$ in $B_2((M\kappa_2(1))^{1/2})$ with $\psi^\varepsilon:=(\varphi^\varepsilon-1)/a(\varepsilon)$. Then,*

$$\mathscr{G}^\varepsilon(\varepsilon N^{\varepsilon^{-1}\varphi_\varepsilon})\Rightarrow\mathscr{G}^0(\psi).$$

Proof By following the lines of argument for Lemma 4.12, to end the proof, it is sufficient to show that $((\overline{Y}^{\varepsilon,\varphi^\varepsilon},\psi^\varepsilon 1_{\{|\psi^\varepsilon|<1/a(\varepsilon)\}}))_{\varepsilon>0}$ is tight in $D([0,T];\mathbb{R}^n)\times B_2((M\kappa_2(1))^{1/2})$. Nevertheless, the tightness of $(\psi^\varepsilon 1_{\{|\psi^\varepsilon|\le1/a(\varepsilon)\}})_{\varepsilon>0}$ follows from [25, Lemma 3.2 (c)] and the weak compactness of $B_2((M\kappa_2(1))^{1/2})$. In what follows, it suffices to demonstrate the tightness of $(\overline{Y}^{\varepsilon,\varphi^\varepsilon})_{\varepsilon>0}$ (see e.g., [24, Theorem A.1]). To achieve this goal, we need to check that $(\overline{Y}^{\varepsilon,\varphi^\varepsilon})_{\varepsilon>0}$ satisfies (I) and (II) in Lemma 4.11. By the Chebyshev inequality, (I) follows from (4.65). Thus it remains to verify (II). Using a Taylor expansion, we obtain from (4.66) that

$$\begin{aligned}\overline{Y}^{\varepsilon,\varphi^\varepsilon}(t)&=I_1^{\varepsilon,\varphi^\varepsilon}(t)+I_4^{\varepsilon,\varphi^\varepsilon}(t)+I_5^{\varepsilon,\varphi^\varepsilon}(t)+\int_0^t\nabla_{\overline{Y}_s^{\varepsilon,\varphi^\varepsilon}}b(X_s^0)\mathrm{d}s\\&\quad+\int_0^t\int_{\mathbb{X}}\nabla_{\overline{Y}_s^{\varepsilon,\varphi^\varepsilon}}\phi(X_s^0,x)\nu(\mathrm{d}x)\mathrm{d}s\\&\quad+a(\varepsilon)\int_0^t\nabla_{\overline{Y}_s^{\varepsilon,\varphi^\varepsilon}}\nabla_{\overline{Y}_s^{\varepsilon,\varphi^\varepsilon}}b(X_s^0+\theta_1(s)(\overline{X}_s^{\varepsilon,\varphi}-X_s^0))\mathrm{d}s\\&\quad+a(\varepsilon)\int_0^t\int_{\mathbb{X}}\nabla_{\overline{Y}_s^{\varepsilon,\varphi^\varepsilon}}\nabla_{\overline{Y}_s^{\varepsilon,\varphi^\varepsilon}}\phi(X_s^0+\theta_2(s)(\overline{X}_s^{\varepsilon,\varphi}-X_s^0),x)\nu(\mathrm{d}x)\mathrm{d}s\\&=:I_1^{\varepsilon,\varphi^\varepsilon}(t)+I_4^{\varepsilon,\varphi^\varepsilon}(t)+I_5^{\varepsilon,\varphi^\varepsilon}(t)+\Theta_1^{\varepsilon,\varphi^\varepsilon}(t)+\Theta_2^{\varepsilon,\varphi^\varepsilon}(t)+R_1^{\varepsilon,\varphi^\varepsilon}(t)+R_2^{\varepsilon,\varphi^\varepsilon}(t)\end{aligned}$$

for some random variables $\theta_1, \theta_2 \in (0,1)$. Hereinafter, let $\{\tau_\varepsilon, \delta_\varepsilon\}$ satisfy (i) and (ii) in Lemma 4.11. Let $\Lambda^{\varepsilon,\varphi^\varepsilon}(t) = I_1^{\varepsilon,\varphi^\varepsilon}(t) + I_4^{\varepsilon,\varphi^\varepsilon}(t) + R_1^{\varepsilon,\varphi^\varepsilon}(t) + R_2^{\varepsilon,\varphi^\varepsilon}(t)$. Since, due to (4.67), (4.68), and (C3),

$$\overline{\mathbb{E}}\Big(\sup_{0\le t\le T}|I_1^{\varepsilon,\varphi^\varepsilon}(t)|^2\Big) + \overline{\mathbb{E}}\Big(\sup_{0\le t\le T}|I_4^{\varepsilon,\varphi^\varepsilon}(t)|^2\Big) + \sum_{i=1}^2 \overline{\mathbb{E}}\Big(\sup_{0\le t\le T}|R_i^{\varepsilon,\varphi^\varepsilon}(t)|^2\Big) \to 0$$

as $\varepsilon \to 0$, it is readily seen that

$$\Lambda^{\varepsilon,\varphi^\varepsilon}(\tau_\varepsilon + \delta_\varepsilon) - \Lambda^{\varepsilon,\varphi^\varepsilon}(\tau_\varepsilon) \to 0 \quad \text{in probability as} \quad \varepsilon \to 0. \tag{4.69}$$

Set $\Xi^{\varepsilon,\varphi^\varepsilon}(t) := I_5^{\varepsilon,\varphi^\varepsilon}(t) + \Theta_1^{\varepsilon,\varphi^\varepsilon}(t) + \Theta_2^{\varepsilon,\varphi^\varepsilon}(t)$. Note from (4.63) that

$$\begin{aligned}
&\overline{\mathbb{E}}|\Xi^{\varepsilon,\varphi^\varepsilon}(\tau_\varepsilon + \delta_\varepsilon) - \Xi^{\varepsilon,\varphi^\varepsilon}(\tau_\varepsilon)| \\
&\le \overline{\mathbb{E}}\int_{\tau_\varepsilon}^{\tau_\varepsilon+\delta_\varepsilon}\int_{\mathbb{X}} |\phi(X_s^0, x))\psi^\varepsilon(s,x)|\nu(\mathrm{d}x)\mathrm{d}s + \overline{\mathbb{E}}\int_{\tau_\varepsilon}^{\tau_\varepsilon+\delta_\varepsilon}\Big|\nabla_{\overline{Y}_s^{\varepsilon,\varphi^\varepsilon}} b(X_s^0)\Big|\mathrm{d}s \\
&\quad + \overline{\mathbb{E}}\int_{\tau_\varepsilon}^{\tau_\varepsilon+\delta_\varepsilon}\int_{\mathbb{X}}\Big|\nabla_{\overline{Y}_s^{\varepsilon,\varphi^\varepsilon}}\phi(X_s^0, x)\Big|\nu(\mathrm{d}x)\mathrm{d}s \\
&\le \Big(1 + \sup_{0\le t\le T}\|X_t^0\|_\infty\Big)\overline{\mathbb{E}}\int_{\tau_\varepsilon}^{\tau_\varepsilon+\delta_\varepsilon}\int_{\mathbb{X}}\Gamma_0(x)\psi^\varepsilon(s,x)|\nu(\mathrm{d}x)\mathrm{d}s \\
&\quad + \Big(1 + \int_{\mathbb{X}}\Gamma_1(x)\nu(\mathrm{d}x)\Big)\overline{\mathbb{E}}\Big(\sup_{0\le t\le T}\|\overline{Y}_t^{\varepsilon,\varphi^\varepsilon}\|_\infty\Big)\delta_\varepsilon \\
&\to 0, \qquad \text{as} \quad \varepsilon \to 0,
\end{aligned}$$

where in the last step we have used [25, Lemma 4.3] and (4.65). So one has

$$\Xi^{\varepsilon,\varphi^\varepsilon}(\tau_\varepsilon + \delta_\varepsilon) - \Xi^{\varepsilon,\varphi^\varepsilon}(\tau_\varepsilon) \to 0 \quad \text{in probability as} \quad \varepsilon \to 0. \tag{4.70}$$

Finally, (II) follows from (4.69) and (4.70) immediately.

Proof of Theorem 4.23

Proof With Lemmas 4.25 and 4.27 in hand, the proof of Theorem 4.23 can be completed.

4.6 Central Limit Theorem for FSDEs of Neutral Type

The MDP scale fills in the gap between the CLT scale and the large deviation scale. For CLTs of SPDEs, we refer to, e.g., [91, 140, 141] and references therein for more details. In this section, we endeavor to investigate a CLT for the following FSDE of neutral type.

$$d\{X^\varepsilon(t) - G(X_t^\varepsilon)\} = b(X_t^\varepsilon)dt + \sqrt{\varepsilon}\sigma(X_t^\varepsilon)dW(t), \quad t > 0, \quad X_0^\varepsilon = \xi \in \mathscr{C}, \quad (4.71)$$

where $\varepsilon \in (0, 1)$, $G, b : \mathscr{C} \mapsto \mathbb{R}^n$, $\sigma : \mathscr{C} \mapsto \mathbb{R}^n \otimes \mathbb{R}^m$. We assume the following conditions hold.

(D1) b is Fréchet differentiable, and there exists an $L > 0$ such that

$$|b(\xi) - b(\eta)| + \|\nabla b(\xi) - \nabla b(\eta)\| + \|\sigma(\xi) - \sigma(\eta)\|_{HS} \leq L\|\xi - \eta\|_\infty, \ \xi, \eta \in \mathscr{C}.$$

(D2) G is Fréchet differentiable, and there exist $\kappa, \kappa_0 \in (0, 1)$ such that

$$|G(\xi) - G(\eta)| \leq \kappa\|\xi - \eta\|_\infty \text{ and } \|\nabla G(\xi) - \nabla G(\eta)\| \leq \kappa_0\|\xi - \eta\|_\infty, \ \xi, \eta \in \mathscr{C}.$$

As $\varepsilon \downarrow 0$, $(X^\varepsilon(t))_{t\geq 0}$, the solution to (4.71), tends to $(X^0(t))_{t\geq 0}$, which solves the following deterministic functional differential equation of neutral type

$$d\{X^0(t) - G(X_t^0)\} = b(X_t^0)dt, \quad t > 0, \quad X_0^0 = \xi \in \mathscr{C}.$$

The CLT concerned with (4.71) is presented as below.

Theorem 4.28 *Under* (D1) *and* (D2),

$$\mathbb{E}\Big(\sup_{0\leq t\leq T}\Big|\frac{X^\varepsilon(t) - X^0(t)}{\sqrt{\varepsilon}} - Y(t)\Big|^2\Big) \lesssim \varepsilon,$$

where $Y(t)$ *solves*

$$d\{Y(t) - \nabla_{Y_t} G(X_t^0)\} = \nabla_{Y_t} b(X_t^0)dt + \sigma(X_t^0)dW(t), \quad t > 0, \quad Y_0 \equiv 0.$$

Proof By the elementary inequality: $(a+b)^p \leq (1+\delta^{\frac{1}{p-1}})^{p-1}(a^p+\delta^{-1}b^p)$, $a, b, \delta > 0, p > 1$, for $M^\varepsilon(t) := X^\varepsilon(t) - G(X_t^\varepsilon)$ we deduce from (D2) that

$$\begin{aligned}|X^\varepsilon(t) - X^0(t)|^4 &\leq (1+\delta^{\frac{1}{3}})^3\{|M^\varepsilon(t) - M^0(t)|^4 + \delta^{-1}|G(X_t^\varepsilon) - G(X_t^0)|^4\}\\ &\leq (1+\delta^{\frac{1}{3}})^3\{|M^\varepsilon(t) - M^0(t)|^4 + \delta^{-1}\kappa^4\|X_t^\varepsilon - X_t^0\|_\infty^4\}.\end{aligned}$$

Taking $\delta = (\kappa/(1-\kappa))^3$ and using $X_0^\varepsilon = X_0^0 = \xi \in \mathscr{C}$ gives that

$$\sup_{0\leq s\leq t}|X^\varepsilon(s) - X^0(s)|^4 \lesssim \sup_{0\leq s\leq t}|M^\varepsilon(s) - M^0(s)|^4.$$

This, together with (D1), Hölder's inequality and B–D–G's inequality, leads to

$$\mathbb{E}\Big(\sup_{0\le s\le t}|X^\varepsilon(s)-X^0(s)|^4\Big)$$
$$\lesssim \int_0^t \mathbb{E}|b(X_s^\varepsilon)-b(X_s^0)|^4\mathrm{d}s+\varepsilon^2\int_0^t\mathbb{E}\|\sigma(X_s^\varepsilon)\|_{HS}^4\mathrm{d}s$$
$$\lesssim \int_0^t \mathbb{E}\Big(\sup_{0\le r\le s}|X^\varepsilon(r)-X^0(r)|^4\Big)\mathrm{d}s+\varepsilon^2\int_0^t(1+\mathbb{E}\|X_s^\varepsilon\|_\infty^4)\mathrm{d}s.$$

So, by virtue of Gronwall's inequality, we obtain from [99, Lemma 4.5, p. 213] that

$$\mathbb{E}\Big(\sup_{0\le s\le t}|X^\varepsilon(s)-X^0(s)|^4\Big)\lesssim\varepsilon^2. \tag{4.72}$$

Set $Y^\varepsilon:=(X^\varepsilon-X^0)/\sqrt{\varepsilon}$. Due to $X_0^\varepsilon=X_0^0=\xi\in\mathscr{C}$ and $Y_0\equiv 0$, one has

$$\begin{aligned}Y^\varepsilon(t)-Y(t)-\nabla_{Y_t^\varepsilon-Y_t}G(X_t^0)&=\frac{1}{\sqrt{\varepsilon}}(G(X_t^\varepsilon)-G(X_t^0))-\nabla_{Y_t^\varepsilon}G(X_t^0)\\&\quad+\int_0^t\Big\{\frac{1}{\sqrt{\varepsilon}}(b(X_s^\varepsilon)-b(X_s^0))-\nabla_{Y_s^\varepsilon}b(X_s^0)\Big\}\mathrm{d}s\\&\quad+\int_0^t\{\sigma(X_s^\varepsilon)-\sigma(X_s^0)\}\mathrm{d}W(s)+\int_0^t\nabla_{Y_s^\varepsilon-Y_s}b(X_s^0)\mathrm{d}s\\&=:\sum_{i=1}^4 I_i(t).\end{aligned}$$

Observe from (D1) and (D2) that

$$|\nabla_\xi\nabla_\xi G(\eta)|\le\kappa_0\|\xi\|_\infty^2\quad\text{and}\quad|\nabla_\xi\nabla_\xi b(\eta)|\le L\|\xi\|_\infty^2,\quad \xi,\eta\in\mathscr{C}. \tag{4.73}$$

By Taylor's expansion theorem, besides Doob's martingale inequality, (D1) and (4.73), we have

$$\begin{aligned}\sum_{i=1}^4\mathbb{E}\Big(\sup_{0\le s\le t}|I_i(t)|^2\Big)&\lesssim\varepsilon\mathbb{E}|\nabla_{Y_t^\varepsilon}\nabla_{Y_t^\varepsilon}G(X_t^0+\theta_1(s)(X_t^\varepsilon-X_t^0))|^2\\&\quad+\varepsilon\int_0^t\mathbb{E}|\nabla_{Y_s^\varepsilon}\nabla_{Y_s^\varepsilon}b(X_t^0+\theta_2(s)(X_t^\varepsilon-X_t^0))|^2\mathrm{d}s\\&\quad+\int_0^t\mathbb{E}\|\sigma(X_s^\varepsilon)-\sigma(X_s^0)\|_{HS}^2\mathrm{d}s\\&\lesssim\varepsilon\int_0^t\{1+\mathbb{E}\|Y_s^\varepsilon\|_\infty^4\}\mathrm{d}s\\&\lesssim\varepsilon\end{aligned} \tag{4.74}$$

for some random variables $\theta_1, \theta_2 \in (0, 1)$, where in the last step we have used (4.72). Next, due to (D1) one has

$$|\nabla_\xi b(\eta)| \le L\|\xi\|_\infty, \quad \xi, \eta \in \mathscr{D}.$$

This, together with (4.74), yields that

$$\mathbb{E}\Big(\sup_{0\le s\le t} |Y^\varepsilon(s) - Y(s) - \nabla_{Y_s^\varepsilon - Y_s} G(X_s^0)|^2\Big) \lesssim \varepsilon + \int_0^t \mathbb{E}\Big(\sup_{0\le r\le s} |Y^\varepsilon(r) - Y(r)|^2\Big)\mathrm{d}s,$$

which, together with $|\nabla_\xi G(\eta)| \le \kappa_0\|\xi\|_\infty$ due to (D2), yields the desired assertion by the elementary inequality and Gronwall's inequality.

Remark 4.29 Theorem 4.28 can be extended to the case of SDDEs, where the coefficients are allowed to bc highly nonlinear w.r.t. the delay variables.

Chapter 5
Stochastic Interest Rate Models with Memory: Long-Term Behavior

In this chapter, we derive the convergence of the long-term return $t^{-\mu}\int_0^t X(s)\mathrm{d}s$ for some $\mu \geq 1$, where X is the short-term interest rate that is an extension of the Cox-Ingersoll-Ross model with jumps and memory. We also investigate the corresponding behavior of the two-factor Cox-Ingersoll-Ross model with jumps and memory. The results of this chapter are based on [16].

5.1 Introduction

Cox et al. [35] proposed a short-term interest rate dynamical system model as

$$\mathrm{d}S(t) = \kappa(\gamma_0 - S(t))\mathrm{d}t + \sigma\sqrt{S(t)}\mathrm{d}W(t)$$

for some $\kappa, \gamma_0, \sigma > 0$. This model is known as the Cox–Ingersoll–Ross (CIR for abbreviation) model and has some empirically relevant properties, e.g., the randomly moving interest rate is elastically pulled toward the long-term constant value γ_0. In order to better capture the properties of empirical data, Chan et al. [29] treated a wide range of models of the short-term interest rate in the framework

$$\mathrm{d}S(t) = \kappa(\gamma_0 - S(t))\mathrm{d}t + \sigma S(t)^{\alpha}\mathrm{d}W(t) \tag{5.1}$$

for $\alpha \geq 1/2$ with appropriate conditions on the parameters κ, γ_0, α. There is an extensive literature on the study of quantitative and qualitative properties of the generalized CIR-type models. For instance, different convergence results and the corresponding applications of the long-term returns can be found in [40, 41, 159]. Strong convergence of the Monte Carlo simulations were studied in [42, 142, 143], and the representations of solutions were presented in [7, 133]. Moreover, the long-term

J. Bao et al., *Asymptotic Analysis for Functional Stochastic Differential Equations*, SpringerBriefs in Mathematics, DOI 10.1007/978-3-319-46979-9_5

returns of the CIR model were investigated in [40, 41], and the results were extended to the jump models in [159].

From an economy point of view, there are some evidences indicating that certain events happened before the trading period influence the current and future asset price, and therefore many scholars introduce delays to the financial models; see, e.g., Arriojas et al. [7], Benhabib [19], and Stoica [133]. Moreover, mean-reverting square root process cannot explain some empirical phenomena like stochastic volatility. To explain these phenomena, jump processes are also used in the financial models; see [18, 30, 66, 106] and references therein.

As described above, there is a natural motivation for considering stochastic interest rate model, where all three features, delay, jump noise, and time dependence of reversion level, are presented. In this chapter, we consider a stochastic interest rate model with jumps and memory in the form

$$\begin{aligned} \mathrm{d}X(t) &= \{2\beta X(t) + \delta(t)\}\mathrm{d}t + \sigma X(t-\tau)^{\gamma}\sqrt{|X(t)|}\mathrm{d}W(t) \\ &\quad + \int_{\Gamma} g(X(t-), u)\tilde{N}(\mathrm{d}t, \mathrm{d}u), \ t > 0 \end{aligned} \tag{5.2}$$

with the initial value $X_0 = \xi \in \mathscr{D}_+$, where $\mathscr{D}_+$ denotes the set of all bounded càdlàg functions $\phi : [-\tau, 0] \mapsto \mathbb{R}_+$. The diffusion term depends on the past through $X^{\gamma}(t-\tau)$ (i.e., delay or memory is involved). Precise assumptions on the parameters of the problem (5.2) are introduced in the next section.

The long-term interest rates play an important role in finance and insurance. For instance, the long-term interest rates determine when homeowners refinance their mortgages in mortgage pricing, play a dominant role in whole-life insurances, and decide when one exchanges a long bond to a short bond in pricing an option. In this light, for instantaneous interest rate model (5.2), it is interesting to investigate the impact of long-term return $t^{-\mu}\int_0^t X(s)\mathrm{d}s$ for some $\mu > 0$. We shall reveal that the long-term return $t^{-\mu}\int_0^t X(s)\mathrm{d}s$ converges almost surely to a stochastic reversion level, which will be stated in Theorem 5.6. As stated in [43], the limit of long-term return $t^{-\mu}\int_0^t X(s)\mathrm{d}s$ is useful in the determination of models of participation in the benefit or of saving products with a guaranteed minimum return.

As is well known, one-factor models imply that the instantaneous returns on bonds of all maturities are perfectly correlated, which is clearly inconsistent with reality (see, e.g., [98]). However, empirical research (see, e.g., [22, 28, 98]) has suggested that two-factor models, which are the short-term interest rate and the instantaneous variance of changes in the short-term interest rate, are better than one-factor models to capture the behavior of the term structure in the real world since the two-factor models allow contingent claim values to reflect both the current level of interest rates as well as the current level of interest rate volatility; see Cox et al. [35, p. 399] and Corzo–Schwartz [34] and references therein.

As an immediate application of Theorem 5.6, we consider the long-term return of the two-factor model in the from

$$\begin{cases} dX(t) = \{2\beta_1 X(t) + \delta(t)\}dt + \sigma_1 X(t-\tau)^{\gamma_1}\sqrt{|X(t)|}dW_1(t) \\ \qquad + \vartheta_1 X(t)\int_\Gamma u\tilde{N}_1(dt, du), \\ dY(t) = \{2\beta_2 Y(t) + X(t)\}dt + \sigma_2 Y(t-\tau)^{\gamma_2}\sqrt{|Y(t)|}dW_2(t) \\ \qquad + \vartheta_2 Y(t)\int_\Gamma u\tilde{N}_2(dt, du) \end{cases} \tag{5.3}$$

with the initial data $(X_0, Y_0) = (\xi, \eta) \in \mathscr{D}_+ \times \mathscr{D}_+$. Here, $(W_1(t))_{t\geq 0}$ and $(W_2(t))_{t\geq 0}$ are Brownian motions, and $\widetilde{N}_1(dt, du) := N_1(dt, du) - \nu_1(du)dt$ and $\widetilde{N}_2(dt, du) := N_2(dt, du) - \nu_2(du)dt$ represent the compensated Poisson measures associated with Poisson counting measures $N_1(dt, du)$ and $N_2(dt, du)$ with the characteristic measures $\lambda_1(\cdot)$ and $\lambda_2(\cdot)$, respectively, defined on $(\Omega, \mathscr{F}, (\mathscr{F}_t)_{t\geq 0}, \mathbb{P})$; $\Gamma \in \mathscr{B}(\mathbb{R}_+)$. More details on the data of model (5.3) are to be presented in the last section. In model (5.3), the short interest rate $Y(t)$ follows an extended CIR model with stochastic reversion level $-X(t)/(2\beta_2)$, where $X(t)$ follows an extension of the square root process. For the model (5.3), we are interested in the almost sure convergence of the long-term return $t^{-\mu}\int_0^t Y(s)ds$ for some $\mu > 0$, which is given in Theorem 5.6. Such convergence is also very useful in finance and insurance market. For example, the customer wants a return as high as possible, and insurance companies wonder how the percentage of interest should be determined when they promise a certain fixed percentage of interest on their insurance products such as bonds, life insurance, and so on.

The rest of this chapter is organized as follows. In Section 5.2, we demonstrate the nonnegative property of $X(t)$ as nominal instantaneous interest rate determined by (5.1), and give a support lemma of Theorem 5.6. Section 5.3 is devoted to the almost sure convergence of long-term return $t^{-\mu}\int_0^t X(s)ds$ for some $\mu > 1/2$ with $X(t)$ being a generalized CIR model determined by (5.2). In the final section, as an application of Theorem 5.6, we consider the almost sure convergence of long-term return $t^{-\mu}\int_0^t Y(s)ds$ for some $\mu > 1/2$ determined by (5.3) with stochastic reversion level $-X(t)/(2\beta_2)$.

5.2 Preliminaries

For the model (5.2), we make the following assumptions.

(A1) $\beta < 0, \sigma > 0$ and $\gamma \in [0, \frac{1}{2})$;
(A2) $\delta : \Omega \times \mathbb{R}_+ \mapsto \mathbb{R}_+$, and there exist constants $\mu > \frac{1}{2}$ and $\nu \geq 0$ such that

$$\lim_{t\to\infty} \frac{1}{t^\mu}\int_0^t \delta(s)ds = \nu \quad \text{a.s.};$$

(A3) $g : \mathbb{R} \times \Gamma \mapsto \mathbb{R}$ and there exists $K > 0$ such that

$$\int_\Gamma |g(x, u) - g(y, u)|^2 \nu(du) \leq K|x - y|^2, \quad x, y \in \mathbb{R};$$

(A4) For any $\theta \in [0, 1]$, $x + \theta g(x, u) \geq 0$ whenever $x > 0$.

Remark 5.1 There are numerous examples such that (A4) holds, e.g., $g(x, u) \geq 0$ for $x \in \mathbb{R}$ and $u \in \Gamma$; $-g(x, u) \leq x$ whenever $g(x, u) \leq 0$.

Since (5.2) is used to model stochastic volatility or interest rate or an asset price, it is critical that the solution $(X(t; \xi))_{t \geq -\tau}$ with the initial value $\xi \in \mathscr{D}_+$ never becomes negative, which is guaranteed by the following lemma.

Lemma 5.2 *Assume that* (A1)–(A4) *hold. Then* (5.2) *has a unique nonnegative solution* $(X(t; \xi))_{t \geq -\tau}$ *for any* $\xi \in \mathscr{D}_+$.

Proof Following the argument of [143, Theorem 4.1], we deduce that the EM approximation sequence $(Y_n(t; \xi))_{n \geq 1}$ associated with (5.2) being Cauchy in the strong L^2 sense and hence converges to a limit, denoted by $X(t; \xi)$, which, in fact, is the unique strong solution to (5.2). In what follows, let $T > 0$ be arbitrary, and write $X(t)$ in lieu of $X(t; \xi)$ for notational simplicity. Moreover, for any $q > 0$, by carrying out the similar argument to [143, Theorem 2.1], there exists $C_{p,T} > 0$ such that

$$\sup_{-\tau \leq t \leq T} \mathbb{E}|X(t)|^q \leq C_{p,T}, \quad q > 0. \tag{5.4}$$

To complete the proof, it remains to show the nonnegative property of the solution $(X(t))_{t \in [0,T]}$. We adopt the method of Yamada–Watanabe [148]. Let $a_0 = 1$ and $a_k = \exp(-k(k+1)/2)$, $k = 1, 2 \ldots$. Then $\int_{a_k}^{a_{k-1}} \frac{1}{kx} \mathrm{d}x = 1$ and there is a continuous nonnegative function $\psi_k(\cdot)$ which enjoys the support (a_k, a_{k-1}), has the integral 1 and satisfies $\psi_k(x) \leq \frac{2}{kx}$. Define an auxiliary function $\phi_k(x) = 0$ for $x \geq 0$ and

$$\phi_k(x) = \int_0^{-x} \mathrm{d}y \int_0^y \psi_k(u)\mathrm{d}u, \quad x < 0.$$

Then, $\phi_k \in C_0^2(\mathbb{R}; \mathbb{R}_+)$ enjoys the following properties:

(i) $-1 \leq \phi_k'(x) \leq 0$ for $-a_{k-1} < x < -a_k$, otherwise $\phi_k'(x) = 0$;
(ii) $|\phi_k''(x)| \leq \frac{2}{k|x|}$ for $-a_{k-1} < x < -a_k$, otherwise $\phi_k''(x) = 0$;
(iii) $x^- - a_{k-1} \leq \phi_k(x) \leq x^-$, $x \in \mathbb{R}$.

Note that $\phi_k(\xi(0)) = 0$ for $\xi \in \mathscr{D}_+$ according to the notion of $\phi_k(\cdot)$, and from (A1), (A2), and (i), that $\beta x \phi'(x) \leq 0$ and $\phi'(x)\delta(\cdot) \leq 0$. So, from (i)–(iii), and Taylor's expansion, in addition to (5.4), it then follows that for $t \in (0, T]$,

$$
\begin{aligned}
\mathbb{E}\phi_k(X(t)) &\leq \frac{\sigma^2}{k}\int_0^t \mathbb{E}|X(s-\tau)|^{2\gamma}\mathrm{d}s + \mathbb{E}\int_0^t\int_\Gamma \{(\phi_k'(X(s)+\theta(s)g(X(s),u)) \\
&\quad - \phi_k'(X(s)))g(X(s),u)\}\nu(\mathrm{d}u)\mathrm{d}s \\
&\leq \frac{\sigma^2 T C_{2\gamma,T}}{k} + \mathbb{E}\int_0^t\int_\Gamma \{(\phi_k'(X(s)+\theta(s)g(X(s),u)) \\
&\quad - \phi_k'(X(s)))g(X(s),u)\}\mathbf{1}_{\{X(s)\leq 0\}}\nu(\mathrm{d}u)\mathrm{d}s \\
&\leq \frac{\sigma^2 T C_{2\gamma,T}}{k} + 2K^{\frac{1}{2}}\lambda^{\frac{1}{2}}(\Gamma)\mathbb{E}\int_0^t X^-(s)\mathbf{1}_{\{X(s)\leq 0\}}\mathrm{d}s \\
&\leq \frac{\sigma^2 T C_{2\gamma,T}}{k} + 2K^{\frac{1}{2}}\lambda^{\frac{1}{2}}(\Gamma)Ta_{k-1} + 2K^{\frac{1}{2}}\lambda^{\frac{1}{2}}(\Gamma)\int_0^t \mathbb{E}\phi_k(X(s))\mathrm{d}s
\end{aligned}
$$

for some random variable $\theta \in [0,1]$, where in the second step we used (i) and (A4), in the third step we utilized (i) and (A3), and in the last step we applied (iii). This, together with the Gronwall inequality, gives that

$$
\mathbb{E}X^-(t) - a_{k-1} \leq \mathbb{E}\phi_k(X(t)) \lesssim k^{-1} + a_{k-1}, \quad t \in (0,T].
$$

Thus, $\mathbb{E}X^-(t) = 0$ as $k \to \infty$ and therefore $X(t) \geq 0$ a.s. for any $t \in (0,T]$. Hence the nonnegative property of the solution $(X(t))_{t\geq 0}$ follows from the arbitrariness of $T > 0$.

We point out that the nonnegativity of $(X(t;\xi))_{t\geq -\tau}$ for any $\xi \in \mathscr{D}_+$ can also be obtained under an alternative condition, which may include the Hölder continuity as an example.

(A4') There exist $\alpha \in [0,1]$ and $\kappa : [0,1] \times \Gamma \mapsto \mathbb{R}_+$ such that

$$
g(x,u)^2\mathbf{1}_{\{-\Phi_\theta(x,u)>0\}} \leq \kappa(\theta,u)|\Phi_\theta(x,u)|^{1+\alpha}\mathbf{1}_{\{-\Phi_\theta(x,u)>0\}}
$$

with

$$
\delta_0 := \sup_{\theta\in[0,1]}\int_\Gamma \kappa(\theta,u)\nu(\mathrm{d}u) < \infty,
$$

where $\Phi_\theta(x,u) := x + \theta g(x,u)$ for any $\theta \in (0,1)$, $x \in \mathbb{R}$ and $u \in \Gamma$.

Remark 5.3 There are a number of examples such that (A4′) holds, e.g., $g(x,u) = \vartheta ux$ for $\vartheta \in \mathbb{R}_+$, $u \in \Gamma \subset \mathscr{B}(\mathbb{R}_+)$ and $x \in \mathbb{R}$. Then, (A4′) holds with $\alpha = 1$, and $\kappa(\theta,u) = \vartheta^2u^2$ whenever $\int_\Gamma u^2\lambda(\mathrm{d}u) < \infty$. Moreover, $g(x,u) = \vartheta ux^{\frac{3}{5}}$ for $\vartheta \in \mathbb{R}_+$, $u \in \Gamma$ and $x \in \mathbb{R}$ also obeys (A4′) with $\alpha = \frac{1}{5}$, and $\kappa(\theta,u) = \vartheta^2u^2$ whenever $\int_\Gamma u^2\lambda(\mathrm{d}u) < \infty$.

The lemma below reveals that the solution to (5.2) enjoys nonnegative property.

Lemma 5.4 *Assume that* (A1)–(A3) *and* (A4′) *hold. Then* (5.2) *has a unique nonnegative solution* $(X(t;\xi))_{t\geq-\tau}$ *for any* $\xi \in \mathscr{D}_+$.

Proof The existence and uniqueness of solution can be obtained by following the arguments of [160, Theorem 4.2] and [143, Theorem 4.1]. For any integer $k \ge 1$, let $\phi_k : \mathbb{R} \mapsto [0, \infty)$ be constructed as follows: $\phi_k(r) = \phi_k'(r) = 0$ for $r \in (-\infty, 0]$, and

$$\phi_k''(r) = \begin{cases} c4k^2 r, & r \in [0, \frac{1}{2k}], \\ -4k^2(r - \frac{1}{k}), & r \in [\frac{1}{2k}, \frac{1}{k}], \\ 0, & \text{otherwise.} \end{cases}$$

By a straightforward calculation, ϕ_k admits an explicit representation

$$\phi_k(r) = \begin{cases} c0, & (-\infty, 0], \\ \frac{2k^2 r^3}{3}, & [0, \frac{1}{2k}], \\ r - \frac{1}{2k} - \frac{2k^2}{3}(r - \frac{1}{k})^3, & [\frac{1}{2k}, \frac{1}{k}], \\ r - \frac{1}{2k}, & [\frac{1}{k}, \infty). \end{cases}$$

Observe that $\phi_k \in C_0^2(\mathbb{R}; \mathbb{R}_+)$ enjoys the following properties: for any $r \in \mathbb{R}$,

(a) $0 \le \phi_k'(r) \le \mathbf{1}_{\{r \ge 0\}}$;
(b) $0 \le \phi_k(r) \uparrow r^+$ as $k \uparrow \infty$;
(c) $\phi_k''(r) \ge 0$ and $r\phi_k''(r) \le \mathbf{1}_{(0,\frac{1}{k})}(r) \downarrow 0$ as $k \uparrow \infty$.

Applying the Itô formula yields that for any $t \in (0, T]$,

$$\begin{aligned} &\mathbb{E}\phi_k(-X(t)) \\ &= \phi_k(-\xi(0)) - \mathbb{E}\int_0^t \phi_k'(-X(s))\delta(s)\mathrm{d}s - 2\beta\mathbb{E}\int_0^t \phi_k'(-X(s))X(s)\mathrm{d}s \\ &\quad + \frac{\sigma^2}{2}\mathbb{E}\int_0^t \phi_k''(-X(s))X(s-\tau)^{2\gamma}X(s)\mathrm{d}s \\ &\quad + \mathbb{E}\int_0^t\int_\Gamma \{\phi_k(-(X(s) + g(X(s), u))) \\ &\quad - \phi_k(-X(s)) + \phi_k'(-X(s))g(X(s), u)\}\lambda(\mathrm{d}u)\mathrm{d}s =: \sum_{i=1}^5 \Xi_i(t). \end{aligned}$$

Because of $\xi \in \mathscr{D}_+$, one has $\Xi_1(t) = 0$ according to the notion of ϕ_k. Note from (a) and (A2) that $\Xi_2(t) \le 0$. Next, observe from (a) that

$$\begin{aligned} \Xi_3(t) &= -2\beta\mathbb{E}\int_0^t \phi_k'(-X(s))X(s)\mathbf{1}_{\{X(s)<0\}}\mathrm{d}s \\ &\le 0. \end{aligned}$$

According to (c), it follows from Hölder's inequality and (5.4) that

$$\begin{aligned}\Xi_4(t) &\lesssim \mathbb{E}\int_0^t X(s-\tau)^{2\gamma}\mathbf{1}_{(0,\frac{1}{k})}(X(s))\mathrm{d}s\\&\lesssim \int_0^t (\mathbb{E}X(s-\tau)^{4\gamma})^{1/2}\Big(\mathbb{E}\mathbf{1}_{(0,\frac{1}{k})}(X(s))\Big)^{1/2}\mathrm{d}s\\&\lesssim \int_0^t \Big(\mathbb{E}\mathbf{1}_{(0,\frac{1}{k})}(X(s))\Big)^{1/2}\mathrm{d}s.\end{aligned}$$

Moreover, employing the Taylor expansion and taking (c) into consideration, we derive that

$$\begin{aligned}\Xi_5(t) &\le \mathbb{E}\int_0^t\int_\Gamma (-\Phi_\theta(X(s),u))\phi_k''(-\Phi_\theta(X(s),u))\\&\quad\times\frac{g(X(s),u)^2}{-\Phi_\theta(X(s),u)}\mathbf{1}_{(0,\frac{1}{k})}(-\Phi_\theta(X(s),u))\lambda(\mathrm{d}u)\mathrm{d}s\\&\le \mathbb{E}\int_0^t\int_\Gamma \frac{g(X(s),u)^2}{-\Phi_\theta(X(s),u)}\mathbf{1}_{(0,\frac{1}{k})}(-\Phi_\theta(X(s),u))\lambda(\mathrm{d}u)\mathrm{d}s\\&\le \frac{1}{\kappa^\alpha}\mathbb{E}\int_0^t\int_\Gamma \kappa(\theta(s),u)\mathbf{1}_{(0,\frac{1}{k})}(-\Phi_\theta(X(s),u))\lambda(\mathrm{d}u)\mathrm{d}s\\&\le \frac{\delta_0 T}{\kappa^\alpha},\end{aligned}$$

where $\Phi_\theta(X(s),u)) = X(s) + \theta(s)g(X(s),u)$ for some random variable $\theta \in [0,1]$. Therefore, we arrive at

$$\mathbb{E}\phi_k(-X(t)) \lesssim \frac{1}{\kappa^\alpha} + \int_0^t \Big(\mathbb{E}\mathbf{1}_{(0,\frac{1}{k})}(X(s))\Big)^{1/2}\mathrm{d}s.$$

Taking $k \uparrow \infty$ followed by recalling (b) and applying Fatou's lemma and the Lebesgue dominated convergence theorem yields

$$\mathbb{E}(-X(t))^+ \le 0, \quad t \in [0,T].$$

Thus, the nonnegative property of $(X(t))_{t\ge-\tau}$ follows from the arbitrariness of T.

The following auxiliary lemma gives an estimate of $(X(t))_{t\ge0}$, determined by the one-factor model (5.2), and plays an important role in analyzing the asymptotic behavior of long-term return $t^{-\mu}\int_0^t X(s)\mathrm{d}s$ for some $\mu \ge 1$.

Lemma 5.5 *Let* (A1)–(A4) *hold and assume further that* $4\beta + K < 0$. *Then there exist* $\kappa, c > 0$ *such that*

$$\mathbb{E}(e^{-\kappa\beta\rho}X^2(\rho)) \lesssim \|\xi\|_\infty^2 + \mathbb{E}\int_0^\rho e^{-\kappa\beta s}(1+\delta^2(s))ds,$$

where $\rho > 0$ *is a bounded stopping time.*

Proof We first recall the Young inequality:

$$a^{\alpha}b^{1-\alpha} \le \alpha a + (1-\alpha)b, \quad a, b > 0, \quad \alpha \in (0,1). \tag{5.5}$$

Let $\kappa, \varepsilon > 0$ be arbitrary. By Itô's formula, (A3) and the inequality (5.5), we obtain

$$\begin{aligned}
\mathrm{d}(e^{-\kappa\beta t}X^2(t)) &= -\kappa\beta e^{-\kappa\beta t}X^2(t)\mathrm{d}t + e^{-\kappa\beta t}\mathrm{d}X^2(t)\\
&= e^{-\kappa\beta t}\Big\{(4-\kappa)\beta X^2(t) + \sigma^2 X(t)X^{2\gamma}(t-\tau) + 2\delta(t)X(t)\\
&\quad + \int_\Gamma g^2(X(t),u)\lambda(\mathrm{d}u)\Big\}\mathrm{d}t + \mathrm{d}M_1(t) + \mathrm{d}M_2(t)\\
&\le e^{-\kappa\beta t}\{((4-\kappa)\beta + \varepsilon + K)X^2(t) + c_1(\varepsilon)X^{4\gamma}(t-\tau)\\
&\quad + c_1(\varepsilon)\delta^2(t)\}\mathrm{d}t + \mathrm{d}M_1(t) + \mathrm{d}M_2(t)\\
&\le e^{-\kappa\beta t}\{((4-\kappa)\beta + \varepsilon + K)X^2(t) + \varepsilon e^{\kappa\beta\tau}X^2(t-\tau)\\
&\quad + c_1(\varepsilon)\delta^2(t) + c_2(\varepsilon)\}\mathrm{d}t + \mathrm{d}M_1(t) + \mathrm{d}M_2(t)
\end{aligned}$$

for some $c_1(\varepsilon), c_2(\varepsilon) > 0$, dependent on ε, where

$$\begin{aligned}
\mathrm{d}M_1(t) &:= 2\sigma e^{-\kappa\beta t}X^{\frac{3}{2}}(t)X^{\gamma}(t-\tau)\mathrm{d}W(t) \qquad \text{and}\\
\mathrm{d}M_2(t) &:= e^{-\kappa\beta t}\int_\Gamma \{g^2(X(t-),u) + 2X(t-)g(X(t-),u)\}\tilde{N}(\mathrm{d}t,\mathrm{d}u).
\end{aligned}$$

Integrating from 0 to ρ and then taking expectations on both sides, we arrive at

$$\begin{aligned}
\mathbb{E}(e^{-\kappa\beta\rho}X^2(\rho)) &\le c\|\xi\|_\infty^2 + ((4-\kappa)\beta + 2\varepsilon + K)\mathbb{E}\int_0^\rho e^{-\kappa\beta s}X^2(s)\mathrm{d}s\\
&\quad + (c_1(\varepsilon)\vee c_2(\varepsilon))\mathbb{E}\int_0^\rho (1+\delta^2(s))\mathrm{d}s.
\end{aligned}$$

As a result, the desired assertion follows by noting that, owing to $4\beta + K < 0$, we can choose $\kappa > 0$ and $\varepsilon > 0$ such that $(4-\kappa)\beta + 2\varepsilon + K = 0$.

5.3 Almost Sure Convergence of Long-Term Returns

Our first main result in this chapter is stated as follows.

Theorem 5.6 *Let* (A1)–(A4) *hold and* $4\beta + K < 0$. *Assume further that there exist* $\lambda > 0$ *and* $\theta \in (0, 2\mu)$ *such that*

$$\limsup_{t\to\infty} \frac{1}{t^\theta}\int_0^t \delta^2(s)ds \le \lambda \ \ a.s. \tag{5.6}$$

Then

$$\lim_{t\to\infty}\frac{1}{t^\mu}\int_0^t\left\{X(s;\xi)+\frac{\delta(s)}{2\beta}\right\}ds=0 \quad a.s. \tag{5.7}$$

for any $\xi\in\mathscr{D}_+$.

Before we present the proof of Theorem 5.6, we cite the following Kronecker's lemma (see, e.g., [121, Exercise 1.16, p. 186] & [159, Lemma 1]).

Lemma 5.7 *Let* $(Y(t))_{t\geq 0}$ *be a càdlàg semimartingale and* f *a continuous strictly positive increasing function which tends to infinity. If* $\int_0^\infty (dY(t)/f(t))$ *exists a.s., then* $Y(t)/f(t)\to 0$ *a.s. as* $t\to\infty$.

Proof of Theorem 5.6 In what follows, we write $(X(t))_{t\geq 0}$ instead of $(X(t;\xi))_{t\geq 0}$. From (5.2), one has

$$\begin{aligned}\int_0^t\left\{X(s)+\frac{\delta(s)}{2\beta}\right\}\mathrm{d}s&=\frac{X(t)-\xi(0)}{2\beta}-\frac{\sigma}{2\beta}\int_0^t X^\gamma(s-\tau)\sqrt{|X(s)|}\mathrm{d}W(s)\\&\quad-\frac{1}{2\beta}\int_0^t\int_\Gamma g(X(s-),u)\tilde{N}(\mathrm{d}s,\mathrm{d}u).\end{aligned} \tag{5.8}$$

On the other hand, an application of Itô's formula to $e^{-2\beta t}X(t)$ yields that

$$\begin{aligned}X(t)&=e^{2\beta t}\Big\{\xi(0)+\int_0^t e^{-2\beta s}\delta(s)\mathrm{d}s+\sigma\int_0^t e^{-2\beta s}X^\gamma(s-\tau)\sqrt{|X(s)|}\mathrm{d}W(s)\\&\quad+\int_0^t\int_\Gamma e^{-2\beta s}g(X(s-),u)\tilde{N}(\mathrm{d}s,\mathrm{d}u)\Big\}.\end{aligned}$$

Thus, substituting this into (5.8) and using Fubini's theorem, we arrive at

$$\begin{aligned}&\frac{1}{t^\mu}\int_0^t\left\{X(s)+\frac{\delta(s)}{2\beta}\right\}\mathrm{d}s\\&=\frac{(e^{2\beta t}-1)\xi(0)}{2\beta t^\mu}+\frac{1}{2\beta t^\mu}\int_0^t e^{2\beta(t-s)}\delta(s)\mathrm{d}s\\&\quad+\frac{\sigma}{2\beta}\Big(1+\frac{1}{t}\Big)^\mu\frac{1}{e^{-2\beta t}(1+t)^\mu}\int_0^t e^{-2\beta s}\delta(s)X^\gamma(s-\tau)\sqrt{|X(s)|}\mathrm{d}W(s)\\&\quad-\frac{\sigma}{2\beta}\Big(1+\frac{1}{t}\Big)^\mu\frac{1}{(1+t)^\mu}\int_0^t X^\gamma(s-\tau)\sqrt{|X(s)|}\mathrm{d}W(s)\\&\quad-\frac{1}{2\beta}\Big(1+\frac{1}{t}\Big)^\mu\frac{1}{(1+t)^\mu}\int_0^t\int_\Gamma g(X(s-),u)\tilde{N}(\mathrm{d}s,\mathrm{d}u)\\&\quad+\frac{1}{2\beta}\Big(1+\frac{1}{t}\Big)^\mu\frac{1}{e^{-2\beta t}(1+t)^\mu}\int_0^t\int_\Gamma e^{-2\beta s}g(X(s-),u)\tilde{N}(\mathrm{d}s,\mathrm{d}u)\\&=:I_1(t)+\frac{1}{2\beta}I_2(t)+\frac{1}{2\beta}\Big(1+\frac{1}{t}\Big)^\mu I_3(t)-\frac{\sigma}{2\beta}\Big(1+\frac{1}{t}\Big)^\mu I_4(t)\\&\quad-\frac{1}{2\beta}\Big(1+\frac{1}{t}\Big)^\mu I_5(t)+\frac{1}{2\beta}\Big(1+\frac{1}{t}\Big)^\mu I_6(t).\end{aligned}$$

To derive assertion (5.7), it is sufficient to verify that $I_i(t) \to 0$ a.s., $i = 1, \ldots, 6$, as $t \to \infty$. Since $\beta < 0$ and $\mu > 0$, $I_1(t) \to 0$ as $t \to \infty$. Due to $\beta < 0$, for any $t \geq 1$ observe that

$$\begin{aligned}
I_2(t) &= \frac{1}{t^\mu}\Big\{\int_0^{t-\sqrt{t}} e^{2\beta\sqrt{t}} e^{2\beta(t-\sqrt{t}-s)}\delta(s)\mathrm{d}s + \int_{t-\sqrt{t}}^t e^{2\beta(t-s)}\delta(s)\mathrm{d}s\Big\}\\
&\leq \frac{e^{2\beta\sqrt{t}}}{t^\mu}\int_0^{t-\sqrt{t}}\delta(s)\mathrm{d}s + \frac{1}{t^\mu}\int_{t-\sqrt{t}}^t \delta(s)\mathrm{d}s\\
&= e^{2\beta\sqrt{t}}\Big(1-\frac{1}{\sqrt{t}}\Big)^\mu \times \frac{1}{(t-\sqrt{t})^\mu}\int_0^{t-\sqrt{t}}\delta(s)\mathrm{d}s + \frac{1}{t^\mu}\int_0^t \delta(s)\mathrm{d}s\\
&\quad - \frac{1}{(t-\sqrt{t})^\mu}\int_0^{t-\sqrt{t}}\delta(s)\mathrm{d}s + \Big(1-\Big(1-\frac{1}{\sqrt{t}}\Big)^\mu\Big)\times\frac{1}{(t-\sqrt{t})^\mu}\int_0^{t-\sqrt{t}}\delta(s)\mathrm{d}s.
\end{aligned}$$

Thus, from (A2) and $\beta < 0$, $I_2(t) \to 0$ a.s. as $t \to \infty$. Next, in order to show $I_3(t) \to 0$ a.s. and $I_4(t) \to 0$ a.s. whenever $t \to \infty$, by Lemma 5.7, it suffices to check that

$$\int_0^\infty \frac{X^\gamma(t-\tau)\sqrt{|X(t)|}}{(1+t)^\mu}\mathrm{d}W(t) \quad \text{exists a.s.} \tag{5.9}$$

For each $n > \|\xi\|_\infty$, define the sequence of stopping times

$$\tau_n = \inf\Big\{t \geq 0 \Big| \int_0^t \frac{\delta^2(s)}{(1+s)^{2\mu}}\mathrm{d}s \geq n\Big\}.$$

In the light of (5.6), there exists an $L > 0$ such that

$$\int_0^t \delta^2(s)ds \leq L(1+t)^\theta \quad \text{a.s.}$$

This, together with $\theta \in (0, 2\mu)$, leads to

$$\begin{aligned}
\int_0^\infty \frac{\delta^2(t)}{(1+t)^{2\mu}}\mathrm{d}t &= \int_0^\infty \frac{1}{(1+t)^{2\mu}}\mathrm{d}\Big(\int_0^t \delta^2(s)\mathrm{d}s\Big)\\
&= \lim_{t\to\infty}\frac{\int_0^t \delta^2(u)\mathrm{d}u}{(1+t)^{2\mu}} + 2\mu\int_0^\infty\Big(\int_0^t \delta^2(u)\mathrm{d}u\Big)\frac{\mathrm{d}t}{(1+t)^{2\mu+1}}\\
&\leq \lim_{t\to\infty}\frac{L}{(1+t)^{2\mu-\theta}} + 2\mu L\int_0^\infty \frac{1}{(1+t)^{2\mu+1-\theta}}\mathrm{d}t\\
&< \infty \quad \text{a.s.}
\end{aligned}$$

Hence, $\{\omega : \tau_n(\omega) = \infty\} \uparrow \Omega$ and therefore it is sufficient to verify (5.9) on $\{\omega : \tau_n(\omega) = \infty\} \uparrow \Omega$. Furthermore, observing that

$$J(t) := \int_0^t \frac{X^\gamma(s-\tau)\sqrt{|X(s)|}}{(1+s)^\mu}\mathbf{1}_{\{s\leq\tau_n\}}\mathrm{d}W(s)$$

is a local martingale, we only need to check that $J(t)$ is an L^2-bounded martingale. By the Itô isometry and inequality (5.5), for $\mu > \frac{1}{2}$ we obtain that

$$\begin{aligned}
\mathbb{E}\,|J(t)|^2 &\leq \int_0^t \frac{\mathbb{E}\{X^2(s)\mathbf{1}_{\{s\leq\tau_n\}}\}}{2(1+s)^{2\mu}}\mathrm{d}s + \int_0^t \frac{\mathbb{E}\{X^{4\gamma}(s-\tau)\mathbf{1}_{\{s\leq\tau_n\}}\}}{2(1+s)^{2\mu}}\mathrm{d}s \\
&\leq \int_0^t \frac{(1-2\gamma)}{2(1+s)^{2\mu}}\mathrm{d}s + \int_0^t \frac{\mathbb{E}\{X^2(s)\mathbf{1}_{\{s\leq\tau_n\}}\}}{2(1+s)^{2\mu}}\mathrm{d}s + \gamma\int_0^t \frac{\mathbb{E}\{X^2(s-\tau)\mathbf{1}_{\{s\leq\tau_n\}}\}}{(1+s)^{2\mu}}\mathrm{d}s \\
&\leq \frac{1-2\gamma}{2(2\mu-1)} + \int_0^t \frac{e^{\kappa\beta s}\mathbb{E}\{e^{-\kappa\beta(s\wedge\tau_n)}X^2(s\wedge\tau_n)\}}{2(1+s)^{2\mu}}\mathrm{d}s \\
&\quad + \gamma\int_0^t \frac{e^{\kappa\beta(s-\tau)}\mathbb{E}\{e^{-\kappa\beta(s\wedge\tau_n-\tau)}X^2(s\wedge\tau_n-\tau)\}}{(1+s)^{2\mu}}\mathrm{d}s \\
&=: (1-2\gamma)/(4\mu-2) + J_1(t) + J_2(t),
\end{aligned}$$

where $\kappa > 0$ is introduced in Lemma 5.5. From Lemma 5.5, in addition to $\mu > \frac{1}{2}$ and $\beta < 0$, it follows that

$$\begin{aligned}
J_1(t) &\lesssim \int_0^t \frac{e^{\kappa\beta s}(e^{-\kappa\beta s}-1) + e^{\kappa\beta s}\mathbb{E}\int_0^{s\wedge\tau_n} e^{-\kappa\beta r}\delta^2(r)\mathrm{d}r}{2(1+s)^{2\mu}}\mathrm{d}s \\
&\lesssim \int_0^t \frac{1 + e^{\kappa\beta s}\int_0^s e^{-\kappa\beta r}\mathbb{E}(\delta^2(r)\mathbf{1}_{\{r\leq\tau_n\}})\mathrm{d}r}{2(1+s)^{2\mu}}\mathrm{d}s \\
&\lesssim 1 + \int_0^t \frac{e^{\kappa\beta s}}{2(1+s)^{2\mu}}\int_0^s e^{-\kappa\beta r}\mathbb{E}(\delta^2(r)\mathbf{1}_{\{r\leq\tau_n\}})\mathrm{d}r\mathrm{d}s \qquad (5.10) \\
&\lesssim 1 + \int_0^t \frac{e^{-\kappa\beta r}}{(1+r)^{2\mu}}\mathbb{E}(\delta^2(r)\mathbf{1}_{\{r\leq\tau_n\}})\int_r^t e^{\kappa\beta s}\mathrm{d}s\mathrm{d}r \\
&\lesssim 1 - \frac{1}{\kappa\beta}\mathbb{E}\int_0^{\tau_n}\frac{\delta^2(r)}{(1+r)^{2\mu}}\mathrm{d}r \\
&\lesssim 1 + n,
\end{aligned}$$

where in the last step we used the notion of stopping time τ_n. For any $t \geq \tau$, due to $\mu > \frac{1}{2}$ and $\beta < 0$, note from Lemma 5.5 that

$$
\begin{aligned}
J_2(t) &\lesssim \|\xi\|_\infty^2 + \int_\tau^t \frac{e^{\kappa\beta(s-\tau)}\mathbb{E}\{e^{-\kappa\beta(s\wedge\tau_n-\tau)}X^2(s\wedge\tau_n-\tau)\}}{(1+s)^{2\mu}}\mathrm{d}s \\
&\lesssim \|\xi\|_\infty^2 + \int_\tau^t \frac{e^{\kappa\beta(s-\tau)}\mathbb{E}\int_0^{s\wedge\tau_n-\tau} e^{-\kappa\beta r}(1+\delta^2(r))\mathrm{d}r}{(1+s)^{2\mu}}\mathrm{d}s \\
&\lesssim \|\xi\|_\infty^2 + \int_\tau^t \frac{1}{(1+s)^{2\mu}}\mathrm{d}s + c\int_\tau^t \frac{e^{\kappa\beta(s-\tau)}\int_0^{s-\tau} e^{-\kappa\beta r}\mathbb{E}(\delta^2(r)\mathbf{1}_{\{r\le\tau_n\}})\mathrm{d}r}{(1+s)^{2\mu}}\mathrm{d}s \\
&\lesssim 1+\|\xi\|_\infty^2 + \int_0^{t-\tau} \frac{e^{-\kappa\beta r}\mathbb{E}(\delta^2(r)\mathbf{1}_{\{r\le\tau_n\}})}{(1+r)^{2\mu}}\int_{r+t}^t e^{\kappa\beta(s-\tau)}\mathrm{d}s\mathrm{d}r \\
&\lesssim 1+\|\xi\|_\infty^2 - \frac{1}{\kappa\beta}\int_0^{t-\tau} \frac{(1-e^{\kappa\beta(t-\tau-r)})\mathbb{E}(\delta^2(r)\mathbf{1}_{\{r\le\tau_n\}})}{(1+r)^{2\mu}}\mathrm{d}r \\
&\lesssim 1+\|\xi\|_\infty^2 - \frac{1}{\kappa\beta}\mathbb{E}\int_0^{\tau_n} \frac{\delta^2(r)}{(1+r)^{2\mu}}\mathrm{d}r \\
&\le 1+n+\|\xi\|_\infty^2.
\end{aligned}
$$

In the above, we also used the definition of stopping time τ_n. Finally, $I_5(t) \to 0$ a.s. and $I_6(t) \to 0$ a.s., respectively, whenever $t \to \infty$ follow by noting from Itô's isometry, and then (A3) and (5.10) that

$$
\begin{aligned}
\mathbb{E}\Big(\int_0^t\int_\Gamma \frac{g(X(s-),u)}{(1+s)^\mu}\mathbf{1}_{\{s\le\tau_n\}}\tilde{N}(\mathrm{d}s,\mathrm{d}u)\Big)^2 &= \mathbb{E}\int_0^t\int_\Gamma \frac{g^2(X(s),u)}{(1+s)^{2\mu}}\mathbf{1}_{\{s\le\tau_n\}}\lambda(\mathrm{d}u)\mathrm{d}s \\
&\le 2KJ_1(t) \\
&\lesssim 1+n.
\end{aligned}
$$

Following the argument of Theorem 5.6 and taking Lemma 5.4 into consideration, we can deduce the following result.

Corollary 5.1 *Let* (A1)–(A3) *and* (5.6) *hold, and assume further that* (A4′) *holds with* $\alpha \in [0, 1)$. *Then, the assertion of Theorem 5.6 still holds.*

Remark 5.8 By Theorem 5.6, under some appropriate conditions on the parameters, we deduce that the long-term return converges almost surely to a reversion level which is random itself. As stated in [40, 41], the long-term returns that determine when homeowners refinance their mortgages in mortgage pricing, play a dominant role in whole-life insurances, and decide when one exchange a long bond to a short bond in pricing an option.

Remark 5.9 For $\beta < 0, \sigma > 0$ and $l \in [0, \frac{1}{2})$, Theorem 5.6 applies to the generalized mean-reverting model

$$
\begin{cases}
\mathrm{d}X(t) = \{2\beta X(t) + \delta(t)\}\mathrm{d}t + \sigma|X(t)|^{\frac{1}{2}+l}\mathrm{d}W(t), \\
X(0) = x > 0,
\end{cases}
$$

whereas Deelstra–Delbaen [40] investigated the long-term returns of such a model only when $l = 0$. Zhao [159] considered the long-time behavior of (5.2) with $\gamma = 0$, $g(x, u) = 0$ for $x < 0$, $u \in U$, and

$$\int_U g^2(x, u)\lambda(\mathrm{d}u) \le K|x| \quad \text{for some constant } K > 0.$$

5.4 An Application to Two-Factor CIR Models

One-factor models imply that the instantaneous returns on bonds of all maturities are perfectly correlated, which is clearly inconsistent with reality (see, e.g., [98]). On the other hand, empirical research (see, e.g., [22, 28, 98]) has suggested that two-factor models, including the short-term interest rate and the instantaneous variance of changes in the short-term interest rate, are better than one-factor models to capture the behavior of the term structure in the real world, because the two-factor models allow contingent claim values to reflect both the current level of interest rates as well as the current level of interest rate volatility.

As an immediate application of Theorem 5.6, in this section, for the two-factor model (5.3), we consider the almost sure convergence of the long-term $t^{-\mu}\int_0^t Y(s)\mathrm{d}s$ for some $\mu > 0$, where the short interest rate $Y(t)$ follows an extended CIR model with stochastic reversion level $-X(t)/(2\beta_2)$, and $X(t)$ follows an extension of the square root process.

For the two-factor model (5.3), we assume that

(A5) $\beta_1 < 0, \sigma_1 > 0, \gamma_1 \in [0, \frac{1}{2}), \vartheta_1 > 0$ and $\delta(\cdot)$ satisfies (A2).
(A6) $\beta_2 < 0, \sigma_2 > 0, \gamma_2 \in [0, \frac{1}{2}), \vartheta_2 > 0$ and $\vartheta_2^2 \int_\Gamma u^2\lambda_2(\mathrm{d}u) < -4\beta_2$.
(A7) For $\theta \in [1, 2\mu)$ (where μ is given in (A2)), $\int_0^\infty \frac{\delta^4(t)}{(1+t)^{2\theta}}\mathrm{d}t < \infty$ a.s.

The following auxiliary lemma gives an estimate of the generalized square root process $(X(t;\xi))_{t\ge 0}$, determined by the first equation in (5.3), and plays a key role in revealing the long-term behavior of two-factor model (5.3).

Lemma 5.10 *Let* (A5) *and* (A6) *hold and assume further that*

$$\Gamma(\vartheta_1, \lambda_1) := \vartheta_1^2 \int_\Gamma u^2(6 + 4\vartheta_1 u + \vartheta_1^2 u^2)\lambda_1(du) < -8\beta_1. \tag{5.11}$$

Then, (5.3) *admits a unique nonnegative solution* $(X(t;\xi), Y(t;\eta))$ *for* $t \ge 0$ *and* $(\xi, \eta) \in \mathscr{D}_+ \times \mathscr{D}_+$, *and there exist* $\kappa > 0$ *and* $c > 0$ *such that*

$$\mathbb{E}(e^{-\kappa\beta_1\rho}X^4(\rho;\xi)) \lesssim 1 + \mathbb{E}\int_0^\rho e^{-\kappa\beta_1 s}(1 + \delta^4(s))ds, \tag{5.12}$$

where $\rho > 0$ *is a bounded stopping time.*

Proof According to Lemma 5.2, under (*A*5) and (*A*6), (5.3) admits a unique nonnegative solution $(X(t;\xi), Y(t;\eta))_{t\geq 0}$ for any $(\xi,\eta)\in\mathscr{D}_+\times\mathscr{D}_+$. For notational simplicity, in what follow, we write $X(t)$ instead of $X(t;\xi)$. Also, by the Itô formula and the elementary inequality (5.5), we derive that

$$\begin{aligned} \mathrm{d}(e^{-\kappa\beta_1 t}X^4(t)) &= -\kappa\beta_1 e^{-\kappa\beta_1 t}X^4(t)\mathrm{d}t + e^{-\kappa\beta_1 t}\mathrm{d}X^4(t) \\ &= e^{-\kappa\beta_1 t}\Big\{(8-\kappa)\beta_1 X^4(t) + 4\delta(t)X^3(t) \\ &\quad + 6\sigma_1^2 X^3(t)X^{2\gamma_1}(t-\tau) + \mathrm{d}\tilde{M}_1(t) + \mathrm{d}\tilde{M}_2(t) \\ &\quad + \int_\Gamma ((1+\vartheta_1 u)^4 - 1 - 4\vartheta_1 u)\lambda_1(\mathrm{d}u)X^4(t)\Big\} \\ &\leq e^{-\kappa\beta_1 t}\{((8-\kappa)\beta_1 + \varepsilon + \Gamma(\vartheta_1,\lambda_1))X^4(t) \\ &\quad + \varepsilon e^{\kappa\beta_1\tau}X^4(t-\tau) + c_\varepsilon(1+\delta^4(t))\}\mathrm{d}t \\ &\quad + \mathrm{d}\tilde{M}_1(t) + \mathrm{d}\tilde{M}_2(t) \end{aligned} \tag{5.13}$$

for any $\kappa > 0$ and sufficiently small $\varepsilon > 0$, where $\tilde{M}_1(t)$ and $\tilde{M}_2(t)$ are two local martingales. Then (5.12) follows by integrating from 0 to ρ, taking expectations on both sides of (5.13) and, in particular, choosing $\kappa > 0$ and $\varepsilon > 0$ sufficiently small such that $(8-\kappa)\beta_1 + 2\varepsilon + \Gamma(\vartheta_1,\lambda_1) = 0$ due to (5.11).

Remark 5.11 In fact, (5.3) admits a unique nonnegative solution $(X(t;\xi), Y(t;\eta))$ for $t \geq 0$ under

$$\Xi(\vartheta_1,\lambda_1) := \vartheta_1^2\int_\Gamma u^2\lambda_1(\mathrm{d}u) < -4\beta_1,$$

rather than (5.11), which is imposed just to guarantee (5.12). Our next result in this chapter presents an almost sure limit.

Theorem 5.12 *Let* (A5)–(A7) *hold, and assume further* (5.6) *and* (5.11) *hold with* $\theta\in[1,2\mu)$. *Then,*

$$\lim_{t\to\infty}\frac{1}{t^\mu}\int_0^t Y(s;\eta)ds = \frac{\nu}{4\beta_1\beta_2} \quad a.s. \tag{5.14}$$

for any $\eta\in\mathscr{D}_+$.

Proof In what follows, we write $(X(t), Y(t))$ in lieu of $(X(t,\xi), X(t,\eta))$ for notation simplicity. According to Theorem 5.6, we deduce from (*A*5) that

$$\lim_{t\to\infty}\frac{1}{t^\mu}\int_0^t X(s)\mathrm{d}s = -\frac{\nu}{2\beta_1} \quad \text{a.s.,} \tag{5.15}$$

where $\nu > 0$ was given in (A2). On the other hand, for $\theta\in[1,2\mu)$ satisfying (5.6), if we can claim that there exists $c > 0$ such that

$$\limsup_{t\to\infty}\frac{1}{t^\theta}\int_0^t X^2(s)\mathrm{d}s \leq c, \quad \text{a.s.} \tag{5.16}$$

Then, by taking (5.15) and Theorem 5.6 into consideration, (5.14) holds. So, it is sufficient to verify that (5.16) holds. Again, by the Itô formula and the elementary inequality (5.5), it follows from (5.3) that

$$\begin{aligned}
\mathrm{d}X^2(t) &= \{(4\beta_1 + m(\vartheta_1, \lambda_1))X^2(t) + 2\delta(t)X(t) + \sigma_1^2 X(t)X^{2\gamma_1}(t-\tau)\}\mathrm{d}t \\
&\quad + 2\sigma_1 X^{\frac{3}{2}}(t)X^{\gamma_1}(t-\tau)\mathrm{d}W_1(t) + \vartheta_1 X^2(t-)\int_\Gamma (2u + \vartheta_1 u^2)\tilde{N}_1(\mathrm{d}u, \mathrm{d}t) \\
&\le \{(4\beta_1 + \varepsilon + m(\vartheta_1, \lambda_1))X^2(t) + \varepsilon X^2(t-\tau) + c_\varepsilon(1 + \delta^2(t))\}\mathrm{d}t \\
&\quad + 2\sigma_1 X^{\frac{3}{2}}(t)X^{\gamma_1}(t-\tau)\mathrm{d}W_1(t) + \vartheta_1 X^2(t-)\int_\Gamma (2u + \vartheta_1 u^2)\tilde{N}_1(\mathrm{d}u, \mathrm{d}t)
\end{aligned}$$

for a sufficiently small $\varepsilon > 0$ and some constant $c_\varepsilon > 0$. Integrating from 0 to t on both sides leads to

$$\begin{aligned}
X^2(t) - \xi^2(0) &\le \varepsilon\|\xi\|_\infty^2 \tau + (4\beta_1 + 2\varepsilon + m(\vartheta_1, \lambda_1))\int_0^t X^2(s)\mathrm{d}s + c_\varepsilon \int_0^t (1 + \delta^2(s))\mathrm{d}s \\
&\quad + 2\sigma_1 \int_0^t X^{\frac{3}{2}}(s)X^{\gamma_1}(s-\tau)\mathrm{d}W_1(s) \\
&\quad + \vartheta_1 \int_0^t \int_\Gamma (2u + \vartheta_1 u^2)X^2(s-)\tilde{N}_1(\mathrm{d}u, \mathrm{d}s).
\end{aligned}$$

By virtue of (5.11), we choose $\varepsilon > 0$ such that $-\tilde{\kappa} := 4\beta_1 + 2\varepsilon + m(\vartheta_1, \lambda_1) < 0$. Thus, for $\theta \in [1, 2\mu)$, it follows that

$$\begin{aligned}
\frac{1}{t^\theta}\int_0^t X^2(s)\mathrm{d}s &\le \frac{c_\varepsilon(\|\xi\|_\infty^2 + t)}{t^\theta} + \frac{c_\varepsilon}{\tilde{\kappa}t^\theta}\int_0^t \delta^2(s)\mathrm{d}s + \frac{2\sigma_1}{\tilde{\kappa}t^\theta}\int_0^t X^{\frac{3}{2}}(s)X^{\gamma_1}(s-\tau)\mathrm{d}W_1(s) \\
&\quad + \frac{\vartheta_1}{\tilde{\kappa}t^\theta}\int_0^t \int_\Gamma (2u + \vartheta_1 u^2)X^2(s-)\tilde{N}_1(\mathrm{d}u, \mathrm{d}s).
\end{aligned}$$

Because of $\theta \in [1, 2\mu)$ and (5.6), it is readily to see that the first two terms on the right-hand side are finite almost surely. In order to prove (5.16), by Lemma 5.7 it suffices to show that

$$J_1(\infty) := \int_0^\infty \frac{X^{\frac{3}{2}}(s)X^{\gamma_1}(s-\tau)}{(1+s)^\theta}\mathrm{d}W_1(s) \quad \text{and}$$
$$J_2(\infty) := \int_0^\infty \int_\Gamma \frac{(u+u^2)X^2(s-)}{(1+s)^\theta}\tilde{N}_1(\mathrm{d}u, \mathrm{d}s)$$

exist a.s., respectively. For each $n > \|\xi\|_\infty$, define the stopping time

$$\rho_n = \inf\left\{t \ge 0 \,\middle|\, \int_0^t \frac{\delta^4(s)}{(1+s)^{2\theta}}\mathrm{d}s \ge n\right\}.$$

By (A7), $\{\omega : \rho_n(\omega) = \infty\} \uparrow \Omega$. So, in what follows, it suffices to show that $J_1(\infty)$ and $J_2(\infty)$ exist, respectively, on $\{\omega : \rho_n(\omega) = \infty\}$. Following the argument of

Theorem 5.6, we need only show that

$$M(t) := \int_0^t \frac{X^{\frac{3}{2}}(s)X^{\gamma_1}(s-\tau)}{(1+s)^\theta}\mathbf{1}_{\{s\le\rho_n\}}\mathrm{d}W_1(s),$$
$$\bar{M}(t) := \int_0^t\int_\Gamma \frac{(u+u^2)X^2(s-)}{(1+s)^\theta}\mathbf{1}_{\{s\le\rho_n\}}\tilde{N}_1(\mathrm{d}u,\mathrm{d}s)$$

are L^2-bounded martingales, respectively. By Itô's isometry and inequality (5.5), for $\theta \ge 1$ it follows that

$$\begin{aligned}\mathbb{E}\,|M(t)|^2 &= \mathbb{E}\int_0^t \frac{X^3(s)X^{2\gamma_1}(s-\tau)}{(1+s)^{2\theta}}\mathbf{1}_{\{s\le\rho_n\}}\mathrm{d}s\\ &\le c+\mathbb{E}\int_0^t \frac{X^4(s)}{(1+s)^{2\theta}}\mathbf{1}_{\{s\le\rho_n\}}\mathrm{d}s+\mathbb{E}\int_0^t \frac{X^4(s-\tau)}{(1+s)^{2\theta}}\mathbf{1}_{\{s\le\rho_n\}}\mathrm{d}s\\ &=: c+\Lambda_1(t)+\Lambda_2(t).\end{aligned}$$

Following the estimates on upper bounds of $J_1(t)$ and $J_2(t)$, respectively, we deduce that

$$\mathbb{E}\,|M(t)|^2 \le c+\Lambda_1(t)+\Lambda_2(t) \lesssim 1+n+\|\xi\|_\infty^2. \tag{5.17}$$

On the other hand, note from (5.11) and (5.17) that

$$\mathbb{E}\,\big|\bar{M}(t)\big|^2 = \int_V (u+u^2)^2\lambda_1(\mathrm{d}u)\Lambda_1(t) \lesssim 1+n+\|\xi\|_\infty^2.$$

The proof is thus complete.

Remark 5.13 From Theorem 5.12, with regard to the two-factor model (5.3), we conclude that the long-term return $t^{-\mu}\int_0^t Y(s)\mathrm{d}s$ converges almost surely to a random variable generally depending on the economic environment. Such convergence is also very useful in finance and insurance market. For example, the customer wants a return as high as possible, and insurance companies wonder how the percentage of interest should be determined when they promise a certain fixed percentage of interest on their insurance products such as bonds, life insurance, and so on.

Remark 5.14 Theorem 5.12 holds for the two-factor CIR-type mode (5.3) with $\tau = 0$ whenever $\gamma_i \in [0, \frac{1}{2})$, $i = 1, 2$. However the model (5.3) is not covered by [159, Theorem 2] due to the fact that the jump-diffusion coefficient is Lipschitz continuous, but not Hölder continuous with exponent $\frac{1}{2}$.

Appendix A
Existence and Uniqueness

To make the brief relatively self-contained, we collect a number of results without proof from the existing literature and put them in two appendices. Although the proofs are omitted, the precise citations of the references are provided.

A.1 Existence and Uniqueness for FSDEs

Consider an n-dimensional FSDE

$$\mathrm{d}X(t) = b(X_t)\mathrm{d}t + \sigma(X_t)\mathrm{d}W(t), \quad t > 0, \quad X_0 = \xi \in \mathscr{C}, \tag{A.1}$$

where $b : \mathscr{C} \mapsto \mathbb{R}^n$ and $\sigma : \mathscr{C} \mapsto \mathbb{R}^n \otimes \mathbb{R}^m$ are Borel measurable, and $(W(t))$ is an m-dimensional Brownian motion defined on $(\Omega, \mathscr{F}, (\mathscr{F}_t)_{t\geq 0}, \mathbb{P})$, a complete filtered probability space.

A $\mathbb{P}$-a.s. continuous $\mathbb{R}^n$-valued and $(\mathscr{F}_t)$-adapted stochastic process $(X(t))_{t\geq -\tau}$ is a strong solution to (A.1) if for all $t \geq 0$

$$X(t) = \xi(0) + \int_0^t b(X_s)\mathrm{d}s + \int_0^t \sigma(X_s)\mathrm{d}W(s), \quad \mathbb{P} - \text{a.s.}$$

We say that the pathwise uniqueness of solutions to (A.1) holds if, for any two solutions $(X(t))$ and $(\bar{X}(t))$ with the same initial value, the two processes $(X(t))$ and $(\bar{X}(t))$ are indistinguishable, namely,

$$\mathbb{P}\{X(t) = \bar{X}(t) \text{ for all } t \geq 0\} = 1.$$

For (A.1), we have the following existence and uniqueness result.

J. Bao et al., *Asymptotic Analysis for Functional Stochastic Differential Equations*, SpringerBriefs in Mathematics, DOI 10.1007/978-3-319-46979-9

Theorem A.1 ([99, p. 150, Theorem 2.2] & [109, p. 36, Theorem 2.1]) *Under the global Lipschitz condition, i.e.,*

$$|b(\xi) - b(\eta)| + \|\sigma(\xi) - \sigma(\eta)\|_{HS} \leq L\|\xi - \eta\|_\infty \tag{A.2}$$

for some constant $L > 0$, (A.1) *has a unique strong solution $(X(t;\xi))_{t\geq-\tau}$ with the initial value $X_0 = \xi \in \mathscr{C}$ such that $\mathbb{E}\Big(\sup_{0\leq t\leq T}\|X_t\|_\infty^p\Big) \leq C_T(1+\|\xi\|_\infty^p)$ for some $C_T > 0$ and any $p > 0$.*

Note that (A.2) implies the linear growth condition, i.e.,

$$|b(\xi)| + \|\sigma(\xi)\|_{HS} \leq L_0(1 + \|\xi\|_\infty), \quad \xi \in \mathscr{C} \tag{A.3}$$

for some $L_0 > 0$. Let B_k be the closed ball in $\mathscr{C}$ with radius k and center 0. We say that b and σ are locally Lipschitzian if for each $k \geq 1$, there is a constant $L_k > 0$ such that

$$|b(\xi) - b(\eta)| + \|\sigma(\xi) - \sigma(\eta)\|_{HS} \leq L_k\|\xi - \eta\|_\infty, \quad \xi, \eta \in B_k. \tag{A.4}$$

If we replace (A.2) by (A.4), together with (A.3), we have the following existence and uniqueness theorem.

Theorem A.2 ([99, p. 154, Theorem 2.6]) *Under* (A.3) *and* (A.4), (A.1) *admits a unique strong solution $(X(t;\xi))_{t\geq-\tau}$ with the initial value $X_0 = \xi \in \mathscr{C}$ such that $\mathbb{E}\Big(\sup_{0\leq t\leq T}\|X_t\|_\infty^p\Big) \leq C_T(1 + \|\xi\|_\infty^p)$ for some $C_T > 0$ and any $p > 0$.*

The local Lipschitz condition (A.4) can be further weakened by the following local weak monotonicity: for each $k \geq 1$, there is a constant $L_k > 0$ such that

$$2\langle\xi(0) - \eta(0), b(\xi) - b(\eta)\rangle + \|\sigma(\xi) - \sigma(\eta)\|_{HS}^2 \leq L_k\|\xi - \eta\|_\infty^2 \tag{A.5}$$

for any $\xi, \eta \in B_k$.

Theorem A.3 ([136, Theorem 2.3]) *Let* (A.5) *hold and assume further that b, σ are bounded on B_k for each $k \geq 1$ and there exists a non-decreasing function $\rho : [0,\infty) \mapsto (0,\infty)$ such that $\int_0^\infty \frac{1}{\rho(u)}du = \infty$ and*

$$2\langle\xi(0), b(\xi)\rangle + \|\sigma(\xi)\|_{HS}^2 \leq \rho(\|\xi\|_\infty^2), \quad \xi \in \mathscr{C}.$$

Then, (A.1) *enjoys a unique strong solution $(X(t;\xi))_{t\geq-\tau}$.*

Consider the following SDDE

$$\mathrm{d}X(t) = b(X(t), X(t-\tau))\mathrm{d}t + \sigma(X(t), X(t-\tau))\mathrm{d}W(t), \quad t > 0 \tag{A.6}$$

with the initial data $X_0 = \xi \in \mathscr{C}$, where $b : \mathbb{R}^n \times \mathbb{R}^n \to \mathbb{R}^n$ and $\sigma : \mathbb{R}^n \times \mathbb{R}^n \to \mathbb{R}^n \otimes \mathbb{R}^m$ are Borel measurable, and $(W(t))$ is an m-dimensional Brownian motion defined on a complete filtered probability space $(\Omega, \mathscr{F}, (\mathscr{F}_t)_{t\geq 0}, \mathbb{P})$. By setting

$$b_0(\xi) = b(\xi(0), \xi(-\tau)) \text{ and } \sigma_0(\xi) = \sigma(\xi(0), \xi(-\tau)), \quad \xi \in \mathscr{C},$$

(A.6) can be rewritten as (A.1).

Let $V : \mathbb{R}^n \times \mathbb{R}^n \mapsto \mathbb{R}_+$ such that

$$V(x, y) \leq K(1 + |x|^q + |y|^q), \quad x, y \in \mathbb{R}^n$$

for some $K, q \geq 1$. For any $x_i, y_i \in \mathbb{R}^n$, $i = 1, 2$, assume that there exists an $L > 0$ such that

$$\begin{aligned} &|b(x_1, y_1) - b(x_2, y_2)| + \|\sigma(x_1, y_1) - \sigma(x_2, y_2)\|_{HS} \\ &\leq L|x_1 - x_2| + V(y_1, y_2)|y_1 - y_2|. \end{aligned} \tag{A.7}$$

Observe that under condition (A.7), the coefficients are allowed to be of nonlinear growth with respect to the delay variables.

Theorem A.4 ([15, Lemma 2.2]) *Under* (A.7), (A.6) *has a unique strong solution* $(X(t;\xi))_{t\geq -\tau}$ *such that* $\mathbb{E}\Big(\sup_{0\leq t\leq T} \|X_t\|_\infty^p\Big) \leq C_T(1 + \|\xi\|_\infty^p)$ *for some* $C_T > 0$ *and any* $p > 0$.

A.2 Existence and Uniqueness for FSDEs of Neutral Type

Consider an n-dimensional FSDE of neutral type

$$\mathrm{d}\{X(t) - G(X_t)\} = b(X_t)\mathrm{d}t + \sigma(X_t)\mathrm{d}W(t), \quad t > 0, \quad X_0 = \xi \in \mathscr{C}. \tag{A.8}$$

Besides (A.5), assume that there exists $\kappa \in (0, 1)$ such that

$$|G(\xi) - G(\eta)| \leq \kappa\|\xi - \eta\|_\infty, \quad \xi, \eta \in \mathscr{C}. \tag{A.9}$$

A $\mathbb{P}$-a.s. continuous $\mathbb{R}^n$-valued $(\mathscr{F}_t)$-adapted stochastic process $(X(t))_{t\geq -\tau}$ is a solution to (A.8) if for all $t \geq 0$

$$X(t) = \xi(0) + G(X_t) - G(\xi) + \int_0^t b(X_s)\mathrm{d}s + \int_0^t \sigma(X_s)\mathrm{d}W(s), \quad \mathbb{P} - \text{a.s.}$$

Theorem A.5 ([99, p. 209, Theorem 2.5]) *Under* (A.5) *and* (A.9), (A.8) *has a unique strong solution* $(X(t;\xi))_{t\geq -\tau}$ *such that* $\mathbb{E}\Big(\sup_{0\leq t\leq T} \|X_t\|_\infty^p\Big) \leq C_T(1 + \|\xi\|_\infty^p)$ *for some* $C_T > 0$ *and any* $p > 0$.

Assume that for each $k \geq 1$, there is a constant $L_k > 0$ such that

$$\begin{aligned} 2\langle \xi(0) - \eta(0) - \{G(\xi) - G(\eta)\}, b(\xi) - b(\eta)\rangle \\ + \|\sigma(\xi) - \sigma(\eta)\|_{HS}^2 \leq L_k \|\xi - \eta\|_\infty^2 \end{aligned} \tag{A.10}$$

for any $\xi, \eta \in B_k$, and that

$$2\langle \xi(0) - G(\xi), b(\xi)\rangle + \|\sigma(\xi)\|_{HS}^2 \leq L(1 + \|\xi\|_\infty^2), \quad \xi \in \mathscr{C} \tag{A.11}$$

for some $L > 0$.

Theorem A.6 ([76, Theorem 2.1]) *Under* (A.9), (A.10), *and* (A.11), (A.8) *has a unique strong solution* $(X(t;\xi))_{t \geq -\tau}$ *such that* $\mathbb{E}\Big(\sup_{0 \leq t \leq T} \|X_t\|_\infty^2\Big) \leq C_T(1 + \|\xi\|_\infty^2)$ *for some* $C_T > 0$.

Consider an SDDE of neutral type

$$\mathrm{d}\{X(t) - G(X(t-\tau))\} = b(X(t), X(t-\tau))\mathrm{d}t + \sigma(X(t), X(t-\tau))\mathrm{d}W(t) \tag{A.12}$$

with the initial value $X_0 = \xi \in \mathscr{C}$, where b, σ, W are defined as in (A.6), and $G : \mathbb{R}^n \mapsto \mathbb{R}^n$. In addition to (A.7), we suppose that

$$|G(x) - G(y)| \leq V(x, y)|x - y|, \quad x, y \in \mathbb{R}^n. \tag{A.13}$$

Theorem A.7 ([75, Lemma 2.2]) *Under* (A.7) *and* (A.13), (A.12) *has a unique strong solution* $(X(t;\xi))_{t \geq -\tau}$ *such that* $\mathbb{E}\Big(\sup_{0 \leq t \leq T} \|X_t\|_\infty^p\Big) \leq C_T(1 + \|\xi\|_\infty^p)$ *for some* $C_T > 0$ *and any* $p > 0$.

A.3 Existence and Uniqueness for FSDEs with Jumps

Let $p_i : D_{p_i} \mapsto \mathbb{R}_0 := \mathbb{R} - \{0\}$ be an adapted process, where D_{p_i} is a countable subset of $\mathbb{R}_+$. Then, as in Ikeda–Watanabe [71, p. 59], for each $i \geq 1$ the Poisson random measure $N_i(\cdot, \cdot) : \mathscr{B}(\mathbb{R}_+ \times \mathbb{R}_0) \times \Omega \mapsto \mathbb{N} \cup \{0\}$, defined on the complete filtered probability space $(\Omega, \mathscr{F}, (\mathscr{F}_t)_{t \geq 0}, \mathbb{P})$, can be represented by

$$N_i((0, t] \times \Gamma) = \sum_{s \in D_p, s \leq t} \mathbf{1}_\Gamma(p_i(s)), \quad \Gamma \in \mathscr{B}(\mathbb{R}_0).$$

In this case, we say that p_i is a Poisson point process and N_i is the Poisson random measure generated by p_i. Let $N_1, \ldots, N_n$ be independent Poisson random measures with the characteristic measures $\nu = (\nu_1, \ldots, \nu_n)$, respectively, and set $N := (N_1, \ldots, N_n)^*$, where v^* denotes the transpose of $v \in \mathbb{R}^n$. We denote by $\widetilde{N}$, the compensated Poisson random measure

$$\widetilde{N}(\mathrm{d}t, \mathrm{d}z) := (N_1(\mathrm{d}t, \mathrm{d}z) - \nu_2(\mathrm{d}z)\mathrm{d}t, \ldots, N_n(\mathrm{d}t, \mathrm{d}z) - \nu_n(\mathrm{d}z)\mathrm{d}t)^*.$$

Let $L^p(\nu) = L^p(\nu, \mathbb{R}^n \otimes \mathbb{R}^n)$, $p \geq 2$, be the collection of measurable functions $f : \mathbb{R}_0 \mapsto \mathbb{R}^n \otimes \mathbb{R}^n$ such that

$$\|f\|_{L^p}^p := \sum_{j=1}^{n}\sum_{i=1}^{n} \|f^{i,j}\|_{L^p(\nu_j,\mathbb{R}^n)}^p < \infty.$$

Consider the following FSDE with jumps

$$\mathrm{d}X(t) = b(X_t)\mathrm{d}t + \sigma(X_t)\mathrm{d}W(t) + \int_{\mathbb{R}_0} \gamma(X_{t-}, z)\tilde{N}(\mathrm{d}t, \mathrm{d}z), \quad t > 0 \tag{A.14}$$

with the initial value $X_0 = \xi \in \mathscr{D}$, where $b : \mathscr{D} \mapsto \mathbb{R}^n$, $\sigma : \mathscr{D} \mapsto \mathbb{R}^n \otimes \mathbb{R}^m$, and $\gamma : \mathscr{D} \times \mathbb{R}_0 \mapsto \mathbb{R}^n \otimes \mathbb{R}^n$.

For any $p \geq 2$, assume that there exists an $L > 0$ such that for any $\xi, \eta \in \mathscr{D}$,

$$\begin{aligned}|b(\xi) - b(\eta)|^p + \|\sigma(\xi) - \sigma(\eta)\|_{HS}^p \\ + \|\gamma(\xi, \cdot) - \gamma(\eta, \cdot)\|_{L^2(\nu)}^p + \|\gamma(\xi, \cdot) - \gamma(\eta, \cdot)\|_{L^p(\nu)}^p \leq L\|\xi - \eta\|_\infty^p,\end{aligned} \tag{A.15}$$

and that

$$|b(\xi)|^p + \|\sigma(\xi)\|_{HS}^p + \|\gamma(\xi, \cdot)\|_{L^2(\nu)}^p + \|\gamma(\xi, \cdot)\|_{L^p(\nu)}^p \leq L(1 + \|\xi\|_\infty^p). \tag{A.16}$$

Theorem A.8 ([8, Theorem 2.8]) *Under* (A.15) *and* (A.16), (A.14) *has a unique solution* $(X(t:\xi))_{t\geq -\tau}$ *such that* $\mathbb{E}\Big(\sup_{0\leq t\leq T} \|X_t(\xi)\|_\infty^p\Big) \leq C_T(1 + \|\xi\|_\infty^p)$ *for some* $C_T > 0$ *and any* $p > 0$.

A.4 Existence and Uniqueness for FSPDEs

Let $(H, \langle\cdot,\cdot\rangle_H, \|\cdot\|_H)$ be a real separable Hilbert space, and $(W(t))_{t\geq 0}$ a cylindrical Wiener process on H under a complete filtered probability space $(\Omega, \mathscr{F}, (\mathscr{F}_t)_{t\geq 0}, \mathbb{P})$; that is,

$$W(t) = \sum_{k=1}^{\infty} B_k(t)e_k, \quad t \geq 0$$

for an orthonormal basis $(e_k)_{k\geq 1}$ on H and a sequence of i.i.d. (independent and identically distributed) $\mathbb{R}$-valued Wiener processes $(B_k(t))_{k\geq 1}$ on $(\Omega, \mathscr{F}, (\mathscr{F}_t)_{t\geq 0}, \mathbb{P})$. Let $(A, \mathscr{D}(A))$ be a negative definite self-adjoint operator on H generating a C_0

contraction semigroup e^{tA}. Let $\mathscr{L}(H)$ and $\mathscr{L}_{HS}(H)$ be the spaces of all bounded linear operators and Hilbert–Schmidt operators on H, respectively. Denote by $\|\cdot\|$ and $\|\cdot\|_{HS}$ the operator norm and the Hilbert–Schmidt norm, respectively. Let $\mathscr{C} = C([-\tau, 0]; H)$, the space of all continuous functions $f : [-\tau, 0] \mapsto H$, equipped with the uniform norm $\|f\|_\infty := \sup_{-\tau\le\theta\le 0} \|f(\theta)\|_H$.

Consider a semi-linear FSPDE on $(H, \langle\cdot,\cdot\rangle_H, \|\cdot\|_H)$ in the form

$$\mathrm{d}X(t) = \{AX(t) + b(X_t)\}\mathrm{d}t + \sigma(X_t)\mathrm{d}W(t), \quad X_0 = \xi \in \mathscr{C}, \tag{A.17}$$

where $b : \mathscr{C} \mapsto H$ and $\sigma : \mathscr{C} \mapsto \mathscr{L}(H)$. A progressively measurable process on H is called a mild solution to (A.17), if for every $t \in [0, T]$,

$$\int_0^t \mathbb{E}\{\|e^{(t-s)A}(b(X_s))\|_H + \|e^{(t-s)A}\sigma(X_s)\|_{HS}^2\}\mathrm{d}s < \infty$$

and almost surely

$$X(t) = e^{tA}\xi(0) + \int_0^t e^{(t-s)A}b(X_s)\mathrm{d}s + \int_0^t e^{(t-s)A}\sigma(X_s)\mathrm{d}W(s).$$

We assume the following conditions hold.

(A1) For each $t > 0$, $e^{tA}b(0) \in H$, and there exists $\varepsilon > 0$ such that

$$\int_0^T \|e^{tA}b(0)\|_H^{2(1+\varepsilon)}\mathrm{d}t < \infty,$$

and there is a positive function $K_b \in C((0, T])$ with $\int_0^t K_b(s)^{1+\varepsilon}\mathrm{d}s < \infty$ such that

$$\|e^{tA}(b(\xi) - b(\eta))\|_H \le K_b(t)\|\xi - \eta\|_\infty^2, \quad t \in [0, T], \quad \xi \in \mathscr{C}.$$

(H2) There exists $\varepsilon > 0$ such that $\int_0^T \|e^{tA}\sigma(0)\|_{HS}^{2(1+\varepsilon)}\mathrm{d}t < \infty$, and there exists a positive function $K_\sigma \in C((0, T])$ such that $\int_0^T K_\sigma(t)^{1+\varepsilon}\mathrm{d}t < \infty$ and

$$\|e^{tA}(\sigma(\xi) - \sigma(\eta))\|_{HS}^2 \le K_\sigma(t)\|\xi - \eta\|_H^2, \quad t \in [0, T], \quad \xi, \eta \in \mathscr{C}.$$

(H3) There exists $\alpha \in (0, 1)$ such that

$$\int_0^T t^{-\alpha}\{K_\sigma(t) + \|e^{tA}\sigma(0)\|_{HS}^2\}\mathrm{d}t < \infty.$$

Theorem A.9 ([138, Theorem 4.1.3]) *Under* (H1)–(H3), (A.17) *admits a unique mild solution* $(X(t;\xi))_{t\geq 0}$ *such that* $\mathbb{E}\Big(\sup_{0\leq t\leq T}\|X_t\|_\infty^{2(1+\varepsilon)}\Big)\leq C_T(1+\|\xi\|_\infty^{2(1+\varepsilon)})$ *for some constant* $C_T>0$.

Appendix B
Markov Property and Variation of Constants Formulas

B.1 Markov Property

Consider (A.1). Note that $X(t) \in \mathbb{R}^n$ and $X_t \in \mathscr{C}$. That is, $X(t)$ is a point in $\mathbb{R}^n$, but X_t is a segment process in $\mathscr{C}$. Since the current state $X(t)$ depends on the history, $X(t)$ is not a Markov process, but the segment process X_t is a Markov process as described as follows.

Theorem B.1 ([109, p. 51, Theorem 3.1.1]) *Under* (A.2), $(X_t)_{t\geq 0}$ *is a Markov process in the sense that*

$$\mathbb{P}(X_t \in B|\mathscr{F}_s) = \mathbb{P}(X_t \in B|X_s), \quad t \geq s$$

for all Borel set $B \in \mathscr{B}(\mathscr{C})$.

Likewise, the segment process associated with (A.14) also enjoys Markov property.

Theorem B.2 [33, Proposition 3.5] & [119, Proposition 3.3] *Under* (A.15) *and* (A.16), $(X_t)_{t\geq 0}$ *is a Markov process in the sense that*

$$\mathbb{P}(X_t \in B|\mathscr{F}_s) = \mathbb{P}(X_t \in B|X_s), \quad t \geq s$$

for all Borel set $B \in \mathscr{B}(\mathscr{D})$.

B.2 Variation of Constants Formulas

Consider the deterministic linear delay differential equation

$$\mathrm{d}X(t) = \Big(\int_{[-\tau,0]} X(t+\theta)\mu(\mathrm{d}\theta)\Big)\mathrm{d}t, \quad t > 0, \quad X_0 = \xi \in \mathscr{C}, \qquad \text{(B.1)}$$

J. Bao et al., *Asymptotic Analysis for Functional Stochastic Differential Equations*, SpringerBriefs in Mathematics, DOI 10.1007/978-3-319-46979-9

where $\mu(\cdot)$ is an $\mathbb{R}^n \otimes \mathbb{R}^n$-valued finite signed measure on $[-\tau, 0]$. The fundamental solution or resolvent of (B.1) is the unique locally absolutely continuous function $\Gamma : [0, \infty) \mapsto \mathbb{R}^n \otimes \mathbb{R}^n$ which satisfies

$$\Gamma(t) = I_{n\times n} + \int_0^t \int_{[\max\{-\tau,-s\},0]} \mu(\mathrm{d}u)\Gamma(s+u)\mathrm{d}s. \tag{B.2}$$

For convenience, we set $\Gamma(t) = \mathbf{0}_{n\times n}$ for $t \in [-\tau, 0)$. The solution $X(\cdot, \xi)$ of (B.1) for an arbitrary initial segment ξ exists, is unique, and can be represented as

$$X(t) = \Gamma(t)\xi(0) + \int_{-\tau}^0 \int_{[-\tau,u]} \Gamma(t+s-u)\mu(\mathrm{d}s)\xi(u)\mathrm{d}u.$$

Define the function $\Delta_\mu : \mathbb{C} \mapsto \mathbb{C}$ by

$$\Delta_\mu(\lambda) = \det\Big(\lambda I_{n\times n} - \int_{[-\tau,0]} e^{\lambda s}\mu(\mathrm{d}s)\Big),$$

where $\det(A)$ signifies the determinant of an $n \times n$ matrix A. Define also the set

$$\Lambda = \{\lambda \in \mathbb{C} : \Delta_\mu(\lambda) = 0\}.$$

The function Δ_ν is analytic, and so the elements of Λ are isolated. Define

$$v_\mu = \sup\{\mathrm{Re}(\lambda) : \Delta_\mu(\lambda) = 0\}.$$

If $v_\mu < 0$, for every $\lambda \in (0, -v_\mu)$, there exists a constant $c_\lambda > 0$ such that

$$\|\Gamma(t)\| \le c_\lambda e^{-\lambda t}, \quad t \ge 0. \tag{B.3}$$

Therefore if $v_\nu < 0$, then Γ decays to zero exponentially. This is a simple restatement of Diekmann et al. [46, 26, Theorem 1.5.4 & Corollary 1.5.5].

Theorem B.3 ([4, Lemma 2.1] & [6, Lemma 1]) *For the fundamental solution Γ solving* (B.2)*, the following are equivalent:*

(1) $v_\mu < 0$;
(2) *Γ decays exponentially as $t \to \infty$;*
(3) $\Gamma \in L^1(\mathbb{R}_+; \mathbb{R}^n \otimes \mathbb{R}^n)$;
(4) $\Gamma \in L^2(\mathbb{R}_+; \mathbb{R}^n \otimes \mathbb{R}^n)$.

Consider the following linear delay differential equation of neutral type

$$\frac{\mathrm{d}}{\mathrm{d}t}\Big\{X(t) - \int_{[-\tau,0]} \rho(\mathrm{d}s)X(t+s)\Big\} = \int_{[-\tau,0]} \mu(\mathrm{d}s)X(t+s), \quad t > 0, \quad X_0 = \xi. \tag{B.4}$$

The fundamental solution or resolvent of (B.4) is the unique continuous function $\Gamma : [0, \infty) \mapsto \mathbb{R}^n \otimes \mathbb{R}^n$ which satisfies

$$\begin{cases} \frac{\mathrm{d}}{\mathrm{d}t}\Big\{\Gamma(t) - \int_{[-\tau,0]} \rho(\mathrm{d}s)\Gamma(t+s)\Big\} = \int_{[-\tau,0]} \mu(\mathrm{d}s)\Gamma(t+s) \\ \Gamma(0) = I_{n\times n}, \quad \Gamma(t) = \mathbf{0}_{n\times n}, \quad t \in [-\tau, 0). \end{cases} \tag{B.5}$$

It plays a role which is analogous to the fundamental system in linear ordinary differential equations and the Green function in partial differential equations. By virtue of [95, Theorem 2.2], $X(t;\xi)$ can be expressed explicitly as

$$\begin{aligned} X(t;\xi) = \Gamma(t)\xi(0) &- \int_{[-\tau,0]} \rho(\mathrm{d}\theta)\Gamma(t+\theta)\xi(0) \\ &+ \int_{[-\tau,0]} \mu(\mathrm{d}\theta)\int_\theta^0 \Gamma(t+\theta-s)\xi(s)\mathrm{d}s \\ &+ \int_{[-\tau,0]} \rho(\mathrm{d}\theta)\int_\theta^0 \Gamma(t+\theta-s)\xi'(s)\mathrm{d}s \end{aligned} \tag{B.6}$$

whenever the initial value $\xi \in W^{1,2}([-\tau,0];\mathbb{R}^n)$. Define the function $\Delta_{\mu,\rho} : \mathbb{C} \mapsto \mathbb{C}$ by

$$\Delta_{\mu,\rho}(\lambda) = \det\Big(\lambda\Big(I_{n\times n} - \int_{[-\tau,0]} e^{\lambda s}\rho(\mathrm{d}s)\Big) - \int_{[-\tau,0]} e^{\lambda s}\mu(\mathrm{d}s)\Big),$$

and

$$v_{\mu,\rho} = \sup\{\mathrm{Re}(\lambda) : \Delta_{\mu,\rho}(\lambda) = 0\}.$$

Theorem B.4 ([6, Lemma 1]) *For the fundamental solution Γ solving* (B.5)*, the following are equivalent:*

(1) $v_{\mu,\rho} < 0$;
(2) *Γ decays exponentially as $t \to \infty$;*
(3) $\Gamma \in L^1(\mathbb{R}_+;\mathbb{R}^n \otimes \mathbb{R}^n)$;
(4) $\Gamma \in L^2(\mathbb{R}_+;\mathbb{R}^n \otimes \mathbb{R}^n)$.

B.3 An Upper Bound of c_λ in (B.3)

For application of Theorem 2.5, we need to estimate the constant c_λ in (B.3). For any $\lambda \in (0, -v_\mu)$, define

$$|||\mu||| = \sup_{1\le k\le n}\sqrt{\sum_{1\le j\le n}\|\mu_{kj}\|_{\mathrm{var}}^2},\ \ T_\lambda = 2e^{\lambda\tau}|||\mu|||,$$

$$\rho_\lambda = \max_{\theta\in[-T_\lambda,T_\lambda]}\left\|\left((\lambda-i\theta)I_{n\times n}+\int_{[-\tau,0]}e^{(-\lambda+i\theta)s}\mu(\mathrm{d}s)\right)^{-1}+(\lambda+\nu_\mu-i\theta)^{-1}I_{n\times n}\right\|.$$

Theorem B.5 *For any* $\lambda\in(0,-\nu_\mu)$,

$$\|\Gamma(t)\|\le\frac{1}{2\pi}\Big\{\frac{(\lambda+\nu_\mu-1)\pi}{\lambda+\nu_\mu}+\frac{4(|\nu_\mu|+e^{\lambda\tau}|||\mu|||)}{T_\lambda}+2\rho_\lambda T_\lambda\Big\}e^{-\lambda t},\ \ t\ge 0.$$

Proof For any $z\ne\nu_\mu$, define

$$Q_z = zI_{n\times n}-\int_{[-\tau,0]}e^{zs}\mu(\mathrm{d}s),\ \ G_z = Q_z^{-1}-\frac{1}{z-\nu_\mu}I_{n\times n}.$$

We have (see [65, Theorem 1.5.1]) that for any $\lambda\in(0,-\nu_\mu)$,

$$\begin{aligned}\Gamma(t)&=\frac{1}{2\pi}\lim_{T\to\infty}\int_{-T}^{T}Q_{-\lambda+i\theta}^{-1}e^{(-\lambda+i\theta)t}\mathrm{d}\theta\\&=\frac{1}{2\pi}\lim_{T\to\infty}\int_{-T}^{T}\Big(G_{-\lambda+i\theta}-\frac{I_{n\times n}}{\lambda+\nu_\mu-i\theta}\Big)e^{(-\lambda+i\theta)t}\mathrm{d}\theta.\end{aligned}\tag{B.7}$$

Note that $\big\|\int_{[-\tau,0]}e^{(-\lambda+i\theta)s}\mu(\mathrm{d}s)\big\|\le e^{\lambda\tau}|||\mu|||$ and

$$\sqrt{1+\lambda^2\theta^{-2}}-\frac{e^{\lambda\tau}|||\mu|||}{|\theta|}\ge\frac{1}{2},\ \ |\theta|\ge T_\lambda.$$

Then

$$\|Q_{-\lambda+i\theta}^{-1}\|\le\frac{1}{\sqrt{\lambda^2+\theta^2}-e^{\lambda\tau}|||\mu|||}\le\frac{2}{|\theta|},\ \ |\theta|\ge T_\lambda.$$

This yields

$$\begin{aligned}\|G_{-\lambda+i\theta}\|&\le\|Q_{-\lambda+i\theta}^{-1}\|\cdot\left\|\frac{\int_{[-\tau,0]}e^{(-\lambda+i\theta)s}\mu(\mathrm{d}s)-\nu_\mu I_{n\times n}}{\lambda+\nu_\mu-i\theta}\right\|\\&\le\frac{2(|\nu_\mu|+e^{\lambda\tau}|||\mu|||)}{|\theta|\sqrt{(\lambda+\nu_\mu)^2+\theta^2}}\le\frac{2(|\nu_\mu|+e^{\lambda\tau}|||\mu|||)}{\theta^2},\ \ |\theta|\ge T_\lambda.\end{aligned}$$

Thus, for any $T\ge T_\lambda$,

$$\begin{aligned}
&\int_{-T}^{T} \|G_{-\lambda+i\theta}e^{(-\lambda+i\theta)t}\|\mathrm{d}\theta \\
&= \int_{-T_\lambda}^{T_\lambda} \|G_{-\lambda+i\theta}e^{(-\lambda+i\theta)t}\|\mathrm{d}\theta + \int_{|\theta|>T_\lambda} \|G_{-\lambda+i\theta}e^{(-\lambda+i\theta)t}\|\mathrm{d}\theta \\
&\le 2\rho_\lambda T_\lambda e^{-\lambda t} + \frac{4(|\nu_\mu| + e^{\lambda\tau}\||\mu|\|)e^{-\lambda t}}{T_\lambda}.
\end{aligned} \tag{B.8}$$

On the other hand,

$$\begin{aligned}
&\lim_{T\to\infty}\int_{-T}^{T} \frac{e^{(-\lambda+i\theta)t}}{\lambda+\nu_\mu - i\theta}\mathrm{d}\theta \\
&= e^{-\lambda t}\lim_{T\to\infty}\int_{-T}^{T}\frac{(\lambda+\nu_\mu)e^{it\theta}}{(\lambda+\nu_\mu)^2+\theta^2}\mathrm{d}\theta + i e^{\lambda t}\lim_{T\to\infty}\int_{-T}^{T}\frac{\theta e^{it\theta}}{(\lambda-\nu_\mu)^2+\theta^2}\mathrm{d}\theta \\
&=: \Theta_1 + \Theta_2.
\end{aligned}$$

It is easy to see that

$$\|\Theta_1\| \le \frac{2e^{-\lambda t}}{-\lambda-\nu_\mu}\lim_{T\to\infty}\arctan\left(\frac{\theta}{-\lambda-\nu_\mu}\right)\Big|_0^T = \frac{\pi e^{-\lambda t}}{-\lambda-\nu_\mu}.$$

Moreover, by the residue theorem,

$$\begin{aligned}
\|\Theta_2\| &= \Big| -2\pi e^{-\lambda t} i\mathrm{Res}\Big[\frac{ze^{itz}}{(-\lambda-\nu_\mu)^2+z^2}, (-\lambda-\nu_\mu)i\Big]\Big| \\
&= \Big| -2\pi e^{-\lambda t} i \lim_{z\to(-\lambda-\nu_\mu)i} (z-(\lambda-\nu_\mu)i)\times\frac{ze^{itz}}{(\lambda+\nu_\mu)^2+z^2}\Big| \\
&= \Big| -2\pi e^{-\lambda t} i \lim_{z\to(-\lambda-\nu_\mu)i}\frac{ze^{itz}}{2(-\lambda-\nu_\mu)i}\Big| \\
&= \Big| -2\pi e^{-\lambda t} i\frac{(-\lambda-\nu_\mu)ie^{t(\lambda+\nu_\mu)}}{2(-\lambda-\nu_\mu)i}\Big| \\
&\le \pi e^{-\lambda t}.
\end{aligned}$$

Hence, we arrive at

$$\left|\lim_{T\to\infty}\int_{-T}^{T}\frac{e^{(-\lambda+i\theta)t}}{\lambda+\nu_\mu-i\theta}\mathrm{d}\theta\right| \le \frac{(\lambda+\nu_\mu-1)\pi e^{-\lambda t}}{\lambda+\nu_\mu}.$$

Combining this with (B.8) and (B.7), the proof is complete.

References

1. Aldous, D. (1978). Stopping times and tightness. *Annals of Probability*, *6*, 335–340.
2. Anh, V., Inoue, A., & Kasahara, Y. (2005). Financial markets with memory II: Innovation processes and expected utility maximization. *Stochastic Analysis and Applications*, *23*, 301–328.
3. Applebaum, D. (2009). *Lévy processes and stochastic calculus* (2nd ed.). Cambridge: Cambridge University Press.
4. Appleby, J. A. D., Mao, X., & Riedle, M. (2009). Geometric Brownian motion with delay: Mean square characterisation. *Proceedings of the American Mathematical Society*, *137*, 339–348.
5. Appleby, J. A. D., Mao, X., & Wu, H. (2010). On the almost sure running maxima of solutions of affine stochastic functional differential equations. *SIAM Journal on Mathematical Analysis*, *42*, 646–678.
6. Appleby, J. A. D., Wu, H., & Mao, X. On the almost sure running maxima of solutions of affine neutral stochastic functional diffenrential equations. arXiv:1310.2349v1.
7. Arriojas, M., Hu, Y., Mohammed, S.-E., & Pap, G. (2007). A delayed Black and Scholes formula. *Stochastic Analysis and Applications*, *25*, 471–492.
8. Bañs, D. R., Cordoni, F., Di Nunno, G., Di Persio, L., & Røse, E. E. Stochastic systems with memory and jumps. arXiv:1603.00272v1.
9. Bao, J., Böttcher, B., Mao, X., & Yuan, C. (2011). Convergence rate of numerical solutions to SFDEs with jumps. *Journal of Applied Mathematics and Computing*, *236*, 119–131.
10. Bao, J., Wang, F.-Y., & Yuan, C. (2013). Transportation cost inequalities for neutral functional stochastic equations. *Zeitschrift fur Analysis und ihre Anwendungen*, *32*, 457–475.
11. Bao, J., Wang, F.-Y., & Yuan, C. (2013). Bismut formulae and applications for functional SPDEs. *Bulletin des Sciences Mathèmatiques*, *137*, 509–522.
12. Bao, J., Yin, G., & Yuan, C. (2014). Ergodicity for functional stochastic differential equations and applications. *Nonlinear Analysis*, *98*, 66–82.
13. Bao, J., Yin, G., & Yuan, C. (2016). Stationary distributions for retarded stochastic differential equations without dissipativity, to appear in *Stochastics*.
14. Bao, J., Yin, G., Yuan, C., & Wang, L. (2014). Exponential ergodicity for retarded stochastic differential equations. *Applied Analysis*, *93*, 2330–2349.
15. Bao, J., & Yuan, C. (2013). Convergence rate of EM scheme for SDDEs. *Proceedings of the American Mathematical Society*, *141*, 3231–3243.
16. Bao, J., & Yuan, C. (2013). Long-term behavior of stochastic interest rate models with jumps and memory. *Insurance: Mathematics & Economics.*, *53*, 266–272.

J. Bao et al., *Asymptotic Analysis for Functional Stochastic Differential Equations*, SpringerBriefs in Mathematics, DOI 10.1007/978-3-319-46979-9

17. Bao, J., & Yuan, C. (2015). Large deviations for neutral functional SDEs with jumps. *Stochastics.*, *87*, 48–70.
18. Bardhan, I., & Chao, X. (1993). Pricing options on securities with discontinuous returns. *Stochastic Processes and their Applications*, *48*, 123–137.
19. Benhabib, J. (2004). Interest rate policy in continuous time with discrete delays. *Journal of Money, Credit and Banking*, *36*, 1–15.
20. Billingsley, P. (1968). *Convergence of probability measures*. New York: Wiley.
21. Bo, L., & Yuan, C. (2016). Stochastic delay differential equations with jump reflection: Invariant measure. *Stochastics*, *88*, 841–863.
22. Brigo, D., & Mercurio, F. (2006). *Interest rate models-theory and practice: With smile, inflation, and credit*. New York: Springer.
23. Budhiraja, A., & Dupuis, P. (2000). A variational representation for positive functionals of infinite dimensional Brownian motion. *Probability and Mathematical Statistics*, *20*, 39–61.
24. Budhiraja, A., Chen, J., & Dupuis, P. (2013). Large deviations for stochastic partial differential equations driven by Poisson random measure. *Stochastic Processes and their Applications*, *123*, 523–560.
25. Budhiraja, A., Dupuis, P., & Ganguly, A. (2016). Moderate deviation principles for stochastic differential equations with jumps. *Annals of Probability*, *44*, 1723–1775.
26. Budhiraja, A., Dupuis, P., & Vasileios, M. (2011). Variational representations for continuous time processes. *Annales de l Institut Henri Poincaré Probabilités et Statistiques*, *47*, 725–747.
27. Butkovsky, O. (2014). Subgeometric rates of convergence of Markov processes in the Wasserstein metric. *Annals of Applied Probability*, *24*, 526–552.
28. Cassola, N., & Barros Luis, J. (2001). *A two-factor model of the German term structure of interest rates*. European Central Bank Working Paper No. 46, available at SSRN: http://ssrn.com/abstract=356020.
29. Chan, K., Karolyi, A., Longstaff, F. A., & Sanders, A. B. (1992). An empirical comparison of alternative models of the short-term interest rate. *Journal of Finance*, *47*, 1209–1227.
30. Chan, T. (1999). Pricing contingent claims on stocks driven by Lévy processes. *Annals of Applied Probability*, *9*, 504–528.
31. Chen, X. (1991). The moderate deviations of independent random vectors in a Banach space. *Chinese Journal of Applied Probability and Statistics*, *7*, 24–33.
32. Cont, R., & Fourni, D.-A. (2013). Functional Itĉalculus and stochastic integral representation of martingales. *Annals of Probability*, *41*, 109–133.
33. Cordonia, F., Di Persiob, L., & Oliv, I. A nonlinear Kolmogorov equation for stochastic functional delay differential equations with jumps. arXiv:1602.03851v1.
34. Corzo, T., & Schwartz, E. (2000). Convergence within the European Union: Evidence from interest rates. *Economic Notes*, *29*, 243–268.
35. Cox, J. C., Ingersoll, J. E., & Ross, S. A. (1985). A theory of the term structure of interest rates. *Econometrica*, *53*, 385–407.
36. Da Prato, G., & Zabczyk, J. (1992). *Stochastic equations in infinite dimensions, encyclopedia of mathematics and its applications*. Cambridge: Cambridge University Press.
37. Da Prato, G., & Zabczyk, J. (1996). *Ergodicity for infinite-dimensional systems*. London Mathematical Society Lecture Note Series (Vol. 229). Cambridge: Cambridge University Press.
38. de Acosta, A. (1992). Moderate deviations and associated Laplace approximations for sums of independent random vectors. *Transactions of the American Mathematical Society*, *329*, 357–375.
39. de Acosta, A. (2000). A general non-convex large deviation result with applications to stochastic equations. *Probability Theory and Related Fields*, *118*, 483–521.
40. Deelstra, G., & Delbaen, F. (1995). Long-term returns in stochastic interest rate models. *Insurance: Mathematics and Economics*, *17*, 163–169.
41. Deelstra, G., & Delbaen, F. (1997). Long-term returns in stochastic interest rate models: Different convergence results. *Applied Stochastic Models and Data Analysis*, *13*, 401–407.

42. Deelstra, G., & Delbaen, F. (1998). Convergence of discretized stochastic (interest rate) processes with stochastic drift term. *Applied Stochastic Models and Data Analysis, 14*, 77–84.
43. Deelstra, G., & Delbaen, F. (2000). Long-term returns in stochastic interest rate models: Application. *Astin Bulletin, 30*, 123–140.
44. Dembo, A., & Zeitouni, O. (1992). *Large deviations techniques and applications*. Boston: Jones and Bartlett Publishers.
45. Deuschel, J.-D., & Stroock, D. W. (1989). *Large deviations*. Boston: Academic Press.
46. Diekmann, O., Gils, S., Lunel, S., & Walther, H. (1995). *Delay equations: Functional-, complex- and nonlinear analysis*. New York: Springer.
47. Dong, Z., Xu, L., & Zhang, X. (2014). Exponential ergodicity of stochastic Burgers equations driven by α-stable processes. *Journal of Statistical Physics, 154*, 929–949.
48. Du, N. H., Dang, N. H., & Dieu, N. T. (2014). On stability in distribution of stochastic differential delay equations with Markovian switching. *Systems & Control Letters, 65*, 43–49.
49. Duncan, T., Pasik-Duncan, B., & Stettner, L. (1994). On the ergodic and the adaptive control of stochastic differential delay systems. *Journal of Optimization Theory and Applications, 81*, 509–531.
50. Dupire, B. Functional Itô's calculus, Bloomberg Portfolio Research Paper No. 2009-04-FRONTIERS, available at SSRN: http://ssrn.com/abstract=1435551.
51. Dupuis, P., & Ellis, R. (1997). *A weak convergence approach to the theory of large deviations*. New York: Wiley.
52. Ekren, I., Touzi, N., & Zhang, J. (2016). Viscosity solutions of fully nonlinear parabolic path dependent PDEs: Part I. *Annals of Probability, 44*, 1212–1253.
53. Elsanousi, I., Øksendal, B., & Sulem, A. (2000). Some solvable stochastic control problems with delay. *Stochastics and Stochastics Reports, 71*, 69–89.
54. Es-Sarhir, A., Scheutzow, M., & van Gaans, O. (2010). Invariant measures for stochastic functional differential equations with superlinear drift term. *Differential and Integral Equations, 23*, 189–200.
55. Ethier, S. N., & Kurtz, T. G. (1986). *Markov processes: Characterization and convergence*. New York: Wiley.
56. Evans, L. C. (1998). *Partial differential equations*. Graduate Studies in Mathematics (Vol. 19). Providence, RI: American Mathematical Society.
57. Fang, S., & Zhang, T. (2005). A study of a class of stochastic differential equations with non-Lipschitzian coefficients. *Probability Theory and Related Fields, 132*, 356–390.
58. Freidlin, M. I., & Wentzell, A. D. (1984). *Random perturbations of dynamical systems*. New York: Springer.
59. Gozlan, N., & Léonard, C. (2007). A large deviation approach to some transportation cost inequalities. *Probability Theory and Related Fields, 139*, 235–283.
60. Gushchin, A., & Küchler, U. (2000). On stationary solutions of delay differential equations driven by a Lévy process. *Stochastic Processes and their Applications, 88*, 195–211.
61. Gyöngy, I. (1998). A note on Euler's approximations. *Potential Analysis, 8*, 205–216.
62. Gyöngy, I., & Rásonyi, M. (2011). A note on Euler approximations for SDEs with Hölder continuous diffusion coefficients. *Stochastic Processes and their Applications, 121*, 2189–2200.
63. Hairer, M. (2008). *Ergodicity for stochastic PDEs*. www.hairer.org/notes/Imperial.pdf.
64. Hairer, M., Mattingly, J. C., & Scheutzow, M. (2011). Asymptotic coupling and a general form of Harris' theorem with applications to stochastic delay equations. *Probability Theory and Related Fields, 149*, 223–259.
65. Hale, J. K., & Lunel, S. M. (1993). *Introduction to functional differential equations*. New York: Springer.
66. Henderson, V., & Hobson, D. (2003). Coupling and option price comparisons in a jump-diffusion model. *Stochastics and Stochastics Reports, 75*, 79–101.
67. Higham, D. J., & Kloeden, P. E. (2005). Numerical methods for nonlinear stochastic differential equations with jumps. *Numerical Mathematics, 101*, 101–119.

68. Higham, D. J., Mao, X., & Stuart, A. M. (2002). Strong convergence of Euler-type methods for nonlinear stochastic differential equations. *SIAM Journal on Numerical Analysis, 40*, 1041–1063.
69. Hu, Y. (1996). Semi-implicit Euler-Maruyama scheme for stiff stochastic equations, The Silvri Workshop, Prog. Probab. 38, H. Koerezlioglu, ed. *Birkhäuser. Boston, 43*, 183–202.
70. Hult, H., & Samorodnitsky, G. (2010). Large deviations for point processes based on stationary sequences with heavy tails. *Journal of Applied Probability, 47*, 1–40.
71. Ikeda, N., & Watanabe, S. (1989). *Stochastic differential equations and diffusion processes*. New York: North-Holland.
72. Itô, K., & Nisio, M. (1964). On stationary solutions of a stochastic differential equation. *Journal of Mathematics of Kyoto University, 4–1*, 1–75.
73. Jacob, N., Wang, Y., & Yuan, C. (2009). Numerical solutions of stochastic differential delay equations with jumps. *Stochastic Analysis and Applications, 27*, 825–853.
74. Jacod, J., & Protter, P. (2003). *Probability essentials* (2nd ed.). Berlin: Spinger.
75. Ji, Y., Bao, J., & Yuan, C. Convergence rate of Euler-Maruyama scheme for SDDEs of neutral type. arXiv:1511.07703.
76. Ji, Y., Song, Q., & Yuan, C. Neutral stochastic differential delay equations with locally monotone coefficients. arXiv:1506.03298.
77. Karatzas, I., & Shreve, S. (1988). *Brownian motion and stochastic calculus, graduate texts in math* (Vol. 113). New York: Springer.
78. Kazmerchuk, Y., Swishchuk, A., & Wu, J. (2007). The pricing of options for securities markets with delayed response. *Mathematics and Computers in Simulation, 75*, 69–79.
79. Kinnally, M. S., & Williams, R. J. (2010). On existence and uniqueness of stationary distributions for stochastic delay differential equations with positivity constraints. *Electronic Journal of Probability, 15*, 409–451.
80. Kloeden, P. E., & Platen, E. (1999). *Numerical solution of stochastic differential equations* (3rd ed.). Berlin: Springer.
81. Kolmanovskii, V. B., & Nosov, V. R. (1981). *Stability and periodic modes of control systems with after effect*. Moscow: Nauka.
82. Kolmogorov, A., & Fomin, S. (1957). *Elements of the theory of functions and functional analysis*. Rochester: Graylock.
83. Komorowski, T., Peszat, S., & Szarek, T. (2010). On ergodicity of some Markov processes. *Annals of Probability, 38*, 1401–1443.
84. Küchler, U., & Platen, E. (2000). Strong discrete time approximation of stochastic differential equations with time delay. *Mathematics and Computers in Simulation, 54*, 189–205.
85. Kushner, H. J. (1968). On the stability of processes defined by stochastic difference-differential equations. *Journal of Differential Equations, 4*, 424–443.
86. Kushner, H. J. (1984). *Approximation and weak convergence methods for random processes with applications to stochastic systems theory*. Cambridge, MA: MIT Press.
87. Kushner, H. J. (2008). *Numerical methods for controlled stochastic delay systems*. Boston: Birkhäuser.
88. Kushner, H. J., & Barnea, D. I. (1970). On the control of a linear functional-differential equation with quadratic cost. *SIAM Journal on Control, 8*, 257–272.
89. Kushner, H. J., & Yin, G. (2003). *Stochastic approximation and recursive algorithms and applications* (2nd ed.). New York: Springer.
90. Lenhart, S., & Travis, C. (1985). Stability of functional partial differential equations. *Journal of Differential Equations, 58*, 212–227.
91. Li, Y., Wang, R., & Zhang, S. (2015). Moderate deviations for a stochastic heat equation with spatially correlated noise. *Acta Applicandae Mathematicae, 139*, 59–80.
92. Liu, K. (2006). *Stability of infinite dimensional stochastic differential equations with applications*. Boca Raton: Chapman & Hall/CRC.
93. Liu, K. (2008). Stochastic retarded evolution equations: Green operators, convolutions, and solutions. *Stochastic Analysis and Applications, 26*, 624–650.

94. Liu, K. (2008). Stationary solutions of retarded Ornstein-Uhlenbeck processes in Hilbert spaces. *Statistics & Probability Letters, 78*, 1775–1783.
95. Liu, K. (2009). The fundamental solution and its role in the optimal control of infinite dimensional neutral systems. *Applied Mathematics and Optimization, 60*, 1–38.
96. Liu, K. (2010). Retarded stationary Ornstein-Uhlenbeck processes driven by Lévy noise and operator self-decomposability. *Potential Analysis, 33*, 291–312.
97. Liu, W. (2010). Large deviations for stochastic evolution equations with small multiplicative noise. *Applied Mathematics and Optimization, 61*, 27–56.
98. Longstaff, F. A., & Schwartz, E. A. (1992). Interest rate volatility and the term structure: A two-factor general equilibrium model. *Journal of Finance, 47*, 1259–1282.
99. Mao, X. (2008). *Stochastic differential equations and applications* (2nd ed.). Chichester: Horwood Publishing Ltd.
100. Mao, X., & Sabanis, S. (2003). Numerical solutions of stochastic differential delay equations under local Lipschitz condition. *Journal of Computational and Applied Mathematics, 151*, 215–227.
101. Marinelli, C., Prévôt, C., & Röckner, M. (2010). Regular dependence on initial data for stochastic evolution equations with multiplicative Poisson noise. *Journal of Functional Analysis, 258*, 616–649.
102. Marinelli, C., & Rökner, M. (2010). Well-posedness and asymptotic behavior for stochastic reaction-diffusion equations with multiplicative Poisson noise. *Electronic Journal of Probability, 15*, 1528–1555.
103. Marinelli, C., & Röckner, M. (2014). *On maximal inequalities for purely discontinuous martingales in infinite dimensions* (pp. 293–315). XLVI: Séminaire de Probabilitiés.
104. Maroulas, V. (2011). Uniform Large deviations for infinite dimensional stochastic systems with jumps. *Mathematika, 57*, 175–192.
105. Mattingly, J. C., Stuart, A. M., & Higham, D. J. (2001). Ergodicity for SDEs and approximations: Locally Lipschitz vector fields and degenerate noise. *Stochastic Processes and their Applications, 101*, 185–232.
106. Merculio, F., & Runggaldier, W. (1993). Option pricing for jump diffusions: Approximations and their interpretation. *Applied Mathematical Finance, 3*, 191–200.
107. Meyn, S., & Tweedie, R. L. (1992). *Stochastic stability of Markov chains*. New York: Springer.
108. Mohamad, S., & Gopalsamy, K. (2000). Continuous and discrete Halanay-type inequalities. *Bulletin of the Australian Mathematical Society, 61*, 371–385.
109. Mohammed, S. (1984). *Stochastic functional differential equations*. Boston: Pitman.
110. Mohammed, S., & Zhang, T. (2006). Large deviations for stochastic systems with memory. *Discrete and Continuous Dynamical Systems-Series B, 6*, 881–893.
111. Nakagiri, S. (1986). Optimal control of linear retarded systems in Banach spaces. *Journal of Mathematical Analysis and Applications, 120*, 169–210.
112. Platen, E., & Bruti-Liberati, N. (2010). *Numerical solution of stochastic differential equations with jumps in finance*. Berlin: Springer.
113. Prévôt, C., & Röckner, M. (2007). *A concise course on stochastic partial differential equations*. Berlin: Springer.
114. Priola, E., Shirikyan, A., Xu, L., & Zabczyk, J. (2012). Exponential ergodicity and regularity for equations with Lévy noise. *Stochastic Processes and their Applications, 122*, 106–133.
115. Priola, E., & Zabczyk, J. (2011). Structural properties of semilinear SPDEs driven by cylindrical stable processes. *Probability Theory and Related Fields, 149*, 97–137.
116. Protter, P. E. (2004). *Stochastic integration and differential equations* (2nd ed.). Berlin: Springer.
117. Reed, M., & Simon, B. (1980). *Methods of modern mathematical physics*. Academic Press.
118. Reiß, M., Riedle, M., & van Gaans, O. (2007). On Émery's inequality and a variation-of-constants formula. *Stochastic Analysis and Applications, 25*, 353–379.
119. Reiß, M., Riedle, M., & van Gaans, O. (2006). Delay differential equations driven by Lévy processes: Stationarity and Feller properties. *Stochastic Processes and their Applications, 116*, 1409–1432.

120. Ren, J., Xu, S., & Zhang, X. (2010). Large deviations for multivalued stochastic differential equations. *Journal of Theoretical Probability*, *23*, 1142–1156.
121. Revuz, D., & Yor, M. (1998). *Continuous Martingales and Brownian motion*. Berlin: Springer.
122. Rey-Bellet, L., & (2006). Ergodic properties of Markov processes, Open quantum systems, II, 1–39 (Vol.,. (1881). *Lecture Notes in Mathematics*. Berlin: Springer.
123. Reynolds, C. W. (1987). Flocks, herds, and schools: A distributed behavioral model. *Computer Graphics*, *21*, 25–34.
124. Röckner, M., & Zhang, T. (2007). Stochastic evolution equations of jump type: Existence, uniqueness and large deviation principles. *Potential Analysis*, *26*, 255–279.
125. Röckner, M., Zhang, T., & Zhang, X. (2010). Large deviations for stochastic tamed 3D Navier-Stokes equations. *Applied Mathematics and Optimization*, *61*, 267–285.
126. Sato, K. (1999). *Lévy processes and infinite divisible distributions*. Cambridge: Cambridge University Press.
127. Scheutzow, M. (1984). Qualitative behaviour of stochastic delay equations with a bounded memory. *Stochastics*, *12*, 41–80.
128. Scheutzow, M. (2013). Exponential growth rate for a singular linear stochastic delay differential equation. *Discrete and Continuous Dynamical Systems*, *18*, 1683–1696.
129. Schurz, H. (1997). *Stability, stationarity, and boundedness of some implicit numerical methods for stochastic differential equations and applications*. Berlin: Logos Verlag Berlin.
130. Situ, R. (2005). *Theory of stochastic differential equations with jumps and applications, mathematical and analytical techniques with applications to engineering*. New York: Springer.
131. Slemrod, M. (1971). Nonexistence of oscillations in a nonlinear distributed network. *Journal of Mathematical Analysis and Applications*, *36*, 22–40.
132. Sowers, R. B., & Zabczyk, J. (2011). Large deviations for stochastic PDEs with Lévy noise. *Journal of Functional Analysis*, *260*, 674–723.
133. Stoica, G. (2004). A stochastic delay financial model. *Proceedings of the American Mathematical Society*, *133*, 1837–1841.
134. Varadhan, S. R. S. (1984). *Large deviations and applications*. Philadelphia: SIAM.
135. Viseck, T., Czirook, A., Ben-Jacob, E., Cohen, I., & Shochet, O. (1995). Novel type of phase transition in a system of self-deriven particles. *Physical Review Letters*, *75*, 1226–1229.
136. von Renesse, M.-K., & Scheutzow, M. (2010). Existence and uniqueness of solutions of stochastic functional differential equations. *Random Operators and Stochastic Equations*, *18*, 267–284.
137. Wang, F.-Y. (2004). *Functional inequalities, Markov semigroups and spectral theory*. Beijing: Science Press.
138. Wang, F.-Y. (2013). *Harnack inequalities for stochastic partial differential equations. Springer Briefs in Mathematics*. New York: Springer.
139. Wang, J. (2012). On the exponential ergodicity of Lévy-driven Ornstein-Uhlenbeck processes. *Journal of Applied Probability*, *49*, 990–1004.
140. Wang, R., Zhai, J., & Zhang, T. S. (2015). A moderate deviation principle for 2-D stochastic Navier-Stokes equations. *Journal of Differential Equations*, *258*, 3363–3390.
141. Wang, R., & Zhang, T. S. (2015). Moderate deviations for stochastic reaction-diffusion equations with multiplicative noise. *Potential Analysis*, *42*, 99–113.
142. Wu, F., Mao, X., & Chen, K. (2008). Strong convergence of Monte Carlo simulations of the mean-reverting square root process with jump. *Applied Mathematics and Computation*, *206*, 494–505.
143. Wu, F., Mao, X., & Chen, K. (2009). The Cox-Ingersoll-Ross model with delay and strong convergence of its Euler-Maruyama approximate solutions. *Applied Numerical Mathematics*, *59*, 2641–2658.
144. Wu, F., Yin, G., & Wang, L. Y. (2012). Stability of a pure random delay system with two-time-scale Markovian switching. *Journal of Differential Equations*, *253*, 878–905.
145. Wu, J. (1996). *Theory and applications of partial functional differential equations*. Berlin: Springer.

146. Wu, J. (2011). Uniform large deviations for multivalued stochastic differential equations with Poisson jumps. *Kyoto Journal of Mathematics*, *51*, 535–559.
147. Wu, L. (1995). Moderate deviations of dependent random variables related to CLT. *Annals of Probability*, *23*, 420–445.
148. Yamada, T., & Watanabe, S. (1971). On the uniqueness of solutions of stochastic differential equations. *Kyoto Journal of Mathematics*, *11*, 155–167.
149. Yang, X., Zhai, J., & Zhang, T. (2015). Large deviations for SPDEs of jump type. *Stochastics and Dynamics*, *15*(1550026), 30.
150. Yin, G., & Ramachandran, K. M. (1990). A differential delay equation with wideband noise perturbation. *Stochastic Processes and their Applications*, *35*, 231–249.
151. Yin, G., Wang, L. Y., & Sun, Y. (2011). Stochastic recursive algorithms for networked systems with delay and random switching: Multiscale formulations and asymptotic properties, SIAM J.: Multiscale Modeling. *Simulation*, *9*, 1087–1112.
152. Yin, G., & Zhang, Q. (2013). *Continuous-time Markov chains and applications: A two-time-scale approach* (2nd ed.). New York, NY: Springer.
153. Yorke, J. A. (1970). Asymptotic stability for one dimensional differential-delay equations. *Journal of Differential Equations*, *7*, 189–202.
154. Yuan, C., & Mao, X. (2008). A note on the rate of convergence of the Euler-Maruyama method for stochastic differential equations. *Stochastic Analysis and Application*, *26*, 325–333.
155. Yuan, C., Zou, J., & Mao, X. (2003). Stability in distribution of stochastic differential delay equations with Markovian switching. *Systems & Control Letters*, *50*, 195–207.
156. Zhang, X. (2008). Euler schemes and large deviations for stochastic Volterra equations with singular kernels. *Journal of Differential Equations*, *244*, 2226–2250.
157. Zhang, X. (2009). Exponential ergodicity of non-Lipschitz stochastic differential equations. *Proceedings of the American Mathematical Society*, *137*, 329–337.
158. Zhang, X. (2013). Derivative formulas and gradient estimates for SDEs driven by α-stable processes. *Stochastic Processes and their Applications*, *123*, 1213–1228.
159. Zhao, J. (2009). Long time behaviour of stochastic interest rate models. *Insurance: Mathematics and Economics*, *44*, 459–463.
160. Zhao, J. (2010). Strong solution of a class of SDEs with jumps. *Stochastic Analysis and Applications*, *5*, 735–746.

Index

J. Bao et al., *Asymptotic Analysis for Functional Stochastic Differential Equations*, SpringerBriefs in Mathematics, DOI 10.1007/978-3-319-46979-9

Printed in the United States
By Bookmasters